# Global Health Informatics

# Global Health Informatics

## How Information Technology Can Change Our Lives in a Globalized World

Edited by

**Heimar de Fátima Marin**

Federal University of São Paulo, São Paulo, Brazil;
Hospital Sirio Libanês, São Paulo, Brazil

**Eduardo Massad**

University of São Paulo, São Paulo, Brazil

**Marco Antonio Gutierrez**

University of São Paulo, São Paulo, Brazil

**Roberto J. Rodrigues**

eHealthStrategies, Bethesda, MD, United States;
Global Health International Advisors, Washington, DC, United States

**Daniel Sigulem**

Federal University of São Paulo — UNIFESP, São Paulo, Brazil

AMSTERDAM • BOSTON • HEIDELBERG • LONDON
NEW YORK • OXFORD • PARIS • SAN DIEGO
SAN FRANCISCO • SINGAPORE • SYDNEY • TOKYO

Academic Press is an imprint of Elsevier

Academic Press is an imprint of Elsevier
125 London Wall, London EC2Y 5AS, United Kingdom
525 B Street, Suite 1800, San Diego, CA 92101-4495, United States
50 Hampshire Street, 5th Floor, Cambridge, MA 02139, United States
The Boulevard, Langford Lane, Kidlington, Oxford OX5 1GB, United Kingdom

British Library Cataloguing-in-Publication Data
A catalogue record for this book is available from the British Library

Library of Congress Cataloging-in-Publication Data
A catalog record for this book is available from the Library of Congress

ISBN: 978-0-12-804591-6

For Information on all Academic Press publications
visit our website at https://www.elsevier.com

Working together
to grow libraries in
developing countries

www.elsevier.com • www.bookaid.org

*Publisher:* Mica Haley
*Acquisition Editor:* Rafael Teixeira
*Editorial Project Manager:* Mariana Kühl Leme
*Production Project Manager:* Chris Wortley
*Designer:* Victoria Pearson

Typeset by MPS Limited, Chennai, India

# Contents

# List of Contributors

**V. Aguiar-Pulido**   Weill Cornell Medicine, New York, NY, United States

**E. Ayres**   NIH Clinical Center, Bethesda, MD, United States

**J.J. Cimino**   University of Alabama at Birmingham, Birmingham, AL, United States

**H. de Fátima Marin**   Federal University of São Paulo, São Paulo, Brazil; Director, Innovation & IT, Hospital Sirio Libanês, São Paulo, Brazil

**R. de Holanda Albuquerque**   Brazilian Diabetes Association, Brazil

**F.G.B. de Quiros**   Hospital Italiano de Buenos Aires, Buenos Aires, Argentina

**P. Degoulet**   Paris Descartes University, Paris, France; INSERM UMR 1138-E22: Information Sciences & Personalized Medicine, Paris, France

**C. Delaney**   American College of Medical Informatics, Bethedsa, MD, United States; University of Minnesota, Minneapolis, MN, United States; American Academy of Nursing, Washington, DC, United States

**W. Ed Hammond**   Duke University School of Medicine, Durham, NC, United States

**C.H. Gattini**   University of Chile, Santiago, Chile

**M.A. Gutierrez**   Heart Institute, Clinics Hospital, University of Sao Paulo Medical School, São Paulo, Brazil

**D. Luna**   Hospital Italiano de Buenos Aires, Buenos Aires, Argentina

**F. Martin-Sanchez**   Weill Cornell Medicine, New York, NY, United States

**E. Massad**   University of São Paulo, São Paulo, Brazil

**R.A. Moreno**   Heart Institute, Clinics Hospital, University of Sao Paulo Medical School, São Paulo, Brazil

**D. Novillo-Ortiz**   Pan American Health Organization/World Health Organization, Washington, DC, United States

**R. Pillay**   University of Alabama at Birmingham, Birmingham, AL, United States

**Y. Quintana**   Beth Israel Deaconess Medical Center, Boston, MA, United States; Harvard Medical School, Boston, MA, United States

**M.P. Ramos**   Federal University of São Paulo — UNIFESP, São Paulo, Brazil

**M.S. Rebelo**   Heart Institute, Clinics Hospital, University of Sao Paulo Medical School, São Paulo, Brazil

**R.J. Rodrigues**   eHealthStrategies, Bethesda, Maryland, United States; Global Health International Advisors, Washington, DC, United States

**C. Safran**   Beth Israel Deaconess Medical Center, Boston, MA, United States; Harvard Medical School, Boston, MA, United States

**D. Sigulem**   Federal University of São Paulo — UNIFESP, São Paulo, Brazil

**C.L. Wen**   University of São Paulo, São Paulo, Brazil

# CHAPTER 1

# Global Health Informatics—An Overview

**Y. Quintana[1,2] and C. Safran[1,2]**
[1]Beth Israel Deaconess Medical Center, Boston, MA, United States
[2]Harvard Medical School, Boston, MA, United States

## INTRODUCTION

Healthcare services are facing growing challenges as a result of rapidly growing populations, people living longer with chronic diseases, advanced treatments involving more healthcare providers, and a limited number of resources to deal with these growing challenges. The World Health Organization has reported that between 2008 and 2030, noncommunicable diseases (WHO, 2008) will dramatically rise worldwide with the burden being the biggest in developing countries (Reardon, 2011). With the advancement of medical science, treatments are becoming more complex, and patients with diseases such as cancer, diabetes, asthma, and cardiovascular diseases are living longer with these conditions and seeing multiple healthcare providers.

The United Nations estimates that the global population aged 60 years and older is expected to more than triple by the year 2050 and will reach 2 billion people (UN, 2015). In 2050, 44% of the world's population will live in countries with at least 20% of the population aged 60 years or older, and one in four people will live in a country where more than 30% of people are above aged 60 years or older. This growing elderly population will require more healthcare services and care coordination.

There will also be a shortage of healthcare providers globally. The global population is growing at a faster rate than the number of available providers who are graduating. From 1970 to 2010 the US physician-to-population ratio increased by 98% (from 161 per 100,000 to 319 per 100,000) (Smart, 2012). Many countries have critical health workforce shortages.

1

Healthcare providers are often not distributed where they are needed most. More than 50% of foreign-born doctors and 40% of foreign-born nurses in the United States are from Asia (Smart, 2012). Numerous countries in the Middle East have a significant shortage of local talent and rely on expatriot communities for both the nursing and physician workforce. In the 2008 WHO/Global Health Workforce Alliance report, the WHO (2008) noted that there is a shortfall of 4.3 million trained healthcare workers globally, with the greatest shortages occurring in the poorest countries. Africa has 10% of the world's population but bears 24% of the global disease burden. It also has 3% of the world's healthcare workforce and less than 1% of the world's financial resources for health. The number of caregivers in 36 countries in Africa is inadequate to deliver even the most basic immunization and maternal health services (Deloitte, 2014). It is estimated that sub-Saharan Africa will need 1.5 million more healthcare workers to provide basic services for its population.

The global migration of patients and healthcare providers is also creating challenges. When patients move from their current city, they often do not carry with them their full patient history to their next healthcare provider. This makes it more challenging for healthcare providers to provide continuity of care for patients and may require duplication of diagnostics tests. Expenditures for global health services are increasing more than 10% in most countries (Deloitte, 2014). There is a global need to find more efficient ways to deliver healthcare services and share medical data while reducing costs and improving outcomes. Moreover, migration creates cultural diversity which presents additional challenges for a healthcare system.

Health informatics can be defined as the acquiring, storing, retrieving, and using of healthcare information to foster better collaboration among a patient's various healthcare providers. Another definition cited by the National Library of Medicine defines health informatics as "the interdisciplinary study of the design, development, adoption and application of IT-based innovations in healthcare services delivery, management, and planning" (NLM, 2016). The term e-health can be defined as "the cost-effective and secure use of information and communications technologies in support of health and health-related fields, including healthcare services, health surveillance, health literature, and health education, knowledge and research" (WHO, 2005). Increasing health informatics is a fundamental requirement for building effective and efficient health information systems at local, national, and global levels (Safran, 2009; McCaffery, 2009).

Other related terms include medical informatics, nursing informatics, clinical informatics, and biomedical informatics (BMI). The scientific study of informatics evaluates approaches to information and knowledge management in clinical care, and public health and biomedical research. The *International Journal of Medical Informatics*, the official journal of the European Federation for Medical Informatics (EFMI) and International Medical Informatics Association (IMIA), describes the field of medical informatics as encompassing the following areas (International Journal of Medical Informatics, 2016):

- Information systems, including national or international registration systems, hospital information systems, departmental and/or physician's office systems, document handling systems, electronic medical record systems, standardization, systems integration, etc.;
- Computer-aided medical decision support systems using heuristic, algorithmic, and/or statistical methods as exemplified in decision theory, protocol development, artificial intelligence, etc.;
- Educational computer-based programs pertaining to medical informatics or medicine in general;
- Organizational, economic, social, and clinical impact, ethical issues, and cost−benefit aspects of IT applications in health care.

BMI is the interdisciplinary field that studies and pursues the effective uses of biomedical data, information, and knowledge for scientific inquiry, problem-solving, and decision-making, motivated by efforts to improve human health. A formal definition of BMI was developed by the American Medical Informatics Association (AMIA) Academic Forum (American Medical Informatics Association, 2016). Subsequently a set of core competencies for BMI were published by the AMIA (Kulikowski et al., 2012).

## GLOBAL HEALTH INFORMATICS

Health informatics systems have been widely developed to support health providers and patients in clinics, hospitals, and at home. In the following we review the goal of these systems, the challenges of their implementation, and evaluations of these systems in developed and developing countries (Blaya et al., 2010).

### Electronic Medical Records

Electronic health record (EHR) systems record health-related information on an individual so that it can be consulted by clinicians or staff for patient care. One formal definition of an EHR is "an electronic version of a patient's medical history, that is maintained by the provider over time, and may include all the key administrative and clinical data relevant to that person's care under a particular provider, including demographics, progress notes, problems, medications, vital signs, past medical history, immunizations, laboratory data, and radiology reports" (CMS, 2016).

The EHR has the potential to streamline the clinician's workflow and to support evidence-based decision support, quality management, and outcomes reporting (Safran et al., 1993; Bates et al., 1998; Kaushal et al., 2003). However, implementation of EHRs can be slow, expensive, and have usability problems (Koppel, 2010; Jamoom and Hing, 2015; Jha, 2011; Kushniruk et al., 2013). In the United States the Department of Veterans Affairs (VA) has developed and deployed the Vista Electronic Health Record system (Evans et al., 2006). Formal evaluations of EHR in developing countries have shown successful implementation. For example, the Indian Health Service's Vista system showed that the majority of clinicians viewed its implementation positively and hence used it more (Sequist et al., 2007). The Mosoriot Medical Record System evaluation in Kenya showed improved staff productivity and reduced patient wait times (Rotich et al., 2003). OPENMRS system is an open source freely available system that has been implemented in Africa (Seebregts et al., 2009), Haiti (Fraser et al., 2004), and Peru (Fraser et al., 2006).

Laboratory information management systems are used to report results to administrators and healthcare personnel. These systems can potentially decrease the time to communicate results, reduce errors, and improve the productivity of a laboratory. These systems have also been deployed in low-resource settings, such as a system in Peru (Blaya et al., 2007), where despite some challenges, such as a limited number of trained personnel, the system was deployed across a wide region.

Pharmacy information systems can be used to order, dispense, or track medications or medication orders, including computerized order entry systems. A systematic review (Robertson et al., 2010) of the impact of computerized pharmacy order entry systems and clinical decision support systems described how usage of these systems resulted in improvements in safety concerns such as drug interactions, contraindications, dose monitoring, and adjustment. These systems can also be used to determine if medication is being prescribed according to clinical guideline recommendations. This study noted that without good communication between pharmacists and physicians, the full benefits of this type of system may not be realized. Developing countries are also deploying these systems. A case−control study of a pharmacy information system in Mexico showed an increase of 41% of patients handled and 28% in the number of tests processed (Alvarez Flores et al., 1995). Training and communication processes are critical issues in the successful implementation of these systems.

## Telehealth Systems

Telehealth systems aim to deliver healthcare services to patients using online, telephone, and text messaging systems. These systems are typically used for: (1) patient care and management such as diagnosis, consultation, and instruction; (2) educational applications including physician education, training for health staff and patient education for both disease management and preventive care, and (3) administration and communication such as exchanging information with insurers, lab services, and scheduling.

Several large-scale telemedicine systems have shown significant outcomes. In the United States, Kaiser Permanente Northern California used a telemedicine system for its 3 million members using patient-friendly Internet, mobile and video tools (Pearl, 2014). The system was used in 2013 for more than 10 million virtual "visits." More than 90% of the healthcare providers indicated that the availability of online tools had allowed them to provide high-quality care for their patients. Also in the United States, between 2003 and 2007, the Veterans Health Administration deployed a national home telehealth program for military service members. An evaluation of 17,025 patients with CCHT showed a 25% reduction in the number of bed days of care, a 19% reduction in the number of hospital admissions, and a mean satisfaction score rating of 86% after enrollment in the program (Darkins et al., 2008).

In the United Kingdom the Whole Systems Demonstrator telehealth system (Henderson, 2014) was developed for patients with diabetes, chronic obstructive pulmonary disease (COPD), and coronary heart disease. The system was implemented to support 6191 patients and 238 GP practices in three cities. An evaluation of 3230 people with diabetes showed that patients in the program had lower mortality and emergency admission rates than the control group (Stevento et al., 2012).

A recent systematic review was conducted to determine the care effectiveness and cost of telehealth interventions in somatic diseases (Elbert, 2014). This report showed that of the 31 articles that were included in the final review, 7 articles (23%) showed that eHealth interventions were effective in either health- or cost-related outcome measures. Another 13 articles (42%) were less confident about the effectiveness or cost-effectiveness of the interventions, and 11 articles (35%) concluded that evidence on efficiency and cost-effectiveness of eHealth interventions was still lacking, limited, or inconsistent. Thus telemedicine systems have had some successful deployments in both developed and developing countries, but to achieve successful outcomes there needs to be careful planning, implementation, and evaluation.

## Mobile Health Systems

Mobile phones provide a new and growing opportunity to expand services to patients. Mobile phones provide an important opportunity to reach large numbers of patients given the rapid rise of phone ownership. Mobile phone applications have been developed for monitoring, evaluating, and tracking patients' status, a clinical decision support system for using a computerized knowledge base, patient medical records, and software algorithms that generate patient-specific recommendations.

A Pew Research Center study (Pew, 2015) showed that 64% of Americans have a smartphone, 85% of Americans aged 19—29 years are smartphone owners, and 62% of smartphone users have used a smartphone to search for health information. The International Telecommunications Union (http://www.itu.int) estimates that more than 2 billion people have mobile phone subscriptions worldwide (ITU, 2015). There are more than 7 billion mobile subscriptions worldwide which is 47% of the population. In 2015, 69% of the global population had 3G mobile broadband, up from 45% in 2011. Four billion people in the developing world remain offline.

An example of a widely used mobile health app is Text4baby (Evans et al., 2012). Text4Baby is an antenatal care mobile health program that delivers text messages to underserved pregnant women and new mothers with the goals of improving their health, healthcare beliefs, practices, and behaviors. A randomized controlled trial involving 90 low-income pregnant women showed improved outcomes of the intervention group that received messages from text4baby in addition to regular healthcare.

There are more than 16,000 health-related apps (Delloitte, 2015). Mobile health app directories maintained by the World Health Organization (WHO, 2016) and Mobile World Capital (MWC, 2016) show a wide range of health applications in all regions of the world. Several projects have demonstrated the feasibility of deploying health services in low- and middle-income countries (Hartzler, 2014). A formal evaluation approach for mobile health systems has also been proposed (Kumar et al., 2013).

## Research Translational Systems

Research informatics systems are used for storing, managing, or reporting on data used for research purposes. These systems can be used for clinical research and basic science research. The data may come from multiple hospitals and from different countries to support collaborative translational research to speed the translation of basic science into clinical practice.

Translatiwas onal Collaboration Platforms (Groves et al., 2013; Biesecker et al., 2012; Canuel et al., 2014) aims to integrate clinical, genomics, and patient-reported data that can be analyzed for biomedical research. These platforms allow clinical researchers, basic science researchers, and data scientists to combine data sets and conduct multidisciplinary research studies. In recent years there has been a significant growth of platforms for translational research. In 2004 caBig (National Cancer Institute, 2015a; Saltz et al., 2006; McConnell et al., 2008) was created by the US National Institutes of Health to support multiinstitutional data sharing and biomedical research. The goals of CaBIG were to: (1) Connect scientists and practitioners through a shareable and interoperable infrastructure, (2) Develop standard rules and a common language to share information more easily, and (3) Build or adapt tools to collect, analyze, integrate, and disseminate information associated with cancer research and care. In 2011 an National Institutes of Health (NIH) study (National Institutes of Health, 2011) reported some of the many problems with the implementation and operation of the caBig program and in May 2012 the program ended (Komatsoulis, 2015). The National Cancer Informatics Program created caGrid as its successor (National Cancer Institute, 2015b).

In 2007 the Informatics for Integrating Biology and the Bedside (i2b2) (National Cancer Institute, 2015c; Murphy et al., 2007) program was funded by the United States NIH to Harvard Medical School Investigators. The i2b2 system can store patient medication information and laboratory values, and these can be combined with clinical research data, such as information from a case report form or genomic data, into a single cohesive unit that can be queried in an integrated manner. The project has grown to be an open-source program and has been adopted by numerous hospitals and research centers around the world for biomedical research.

In 2003 The Pediatric Oncology Network Database (www.pond4kids.org) (Quintana et al., 2013) was launched as a program of the International Outreach Program at St. Jude Children's Research Hospital. POND4Kids is a secure, web-based, multilingual pediatric hematology/oncology database created for use in countries with limited resources. The system is used for cancer registration, tracking protocol-based cancer care, outcome evaluation, and assessment and comparison of results among centers. Cancer centers used it to achieve uniform data collection to facilitate meaningful comparison of the implementation of established cancer care protocols.

## Training Programs

Informatics training programs will be key to the development of the informatics field for both industry professionals and academic fields (Detmer 2014). Several groups have outlined the training needs for industry professionals and developed training curricula (NORC, 2014; Mohla, 2014). Nursing informatics training programs have also been developed. The TIGER program is one of the most extensive nursing informatics programs (McCormick et al., 2007, Sipes, 2016). This program has been translated into multiple languages (Kuo et al., 2012). International standards for academic informatics education have been developed by Hasman and Mantas (2013) and Mantas et al. (2013). These programs have been formally evaluated in both developed countries (Mihalas et al., 2014) and developing countries (Luna et al., 2014).

## Challenges

The growing range of informatics systems supports improvements in health-care services, patient support programs, medical research, and collaborative translational research. Collaborative networks are forming worldwide, creating opportunities for larger data sets and more in-depth analysis, as well as comparison of outcomes. However, there are many challenges to establishing and maintaining these systems. We outline a few of the major issues below.

- Technical Data Integration: The growing volume and complexity of clinical and research data require more complex architectures to integrate data from diverse data sets. Data from different sources is difficult because of different data formats.
- Privacy and Regulation: Each country and region will have local privacy laws making data sharing both technically and legally complex. Systems may need to have audit trails of data access to meet regulatory compliance requirements.
- Aging Population: The aging population that will result in more elderly healthcare needs and multiple healthcare service providers. Many patients may prefer home care, or their hospitals may prefer them to receive home care to reduce costs and improve outcomes. This is an opportunity to develop new care delivery models, but the design of these systems needs to take into account elder preferences (Crotty et al., 2015).
- Mobile Technology: Mobile technology is becoming cheaper and more widely available. However, the field of mobile health apps is so new that there are few providers that have trained staff to launch, manage, and integrate data that comes from mobile phones. We may also have a case where we collect more data from mobile phones than we have time and resources to analyze that data and integrate it into care plans for patients.
- Accessibility: Internet and phone accessibility is increasing worldwide. As more people have access to both the Internet and mobile phones, more patients will be able to communicate with healthcare providers from a distance with healthcare providers. We need to ensure that services are provided on multiple communication channels.
- Adaptation: Many systems particularly EMR systems are based on local billing and regulatory requirements. To make these systems more globally available, they need to be more flexible to adapt to global customizations that take into account different workflows and user interfaces.
- Implementation: These systems require careful planning, scope definition, and implementation strategies. Without careful planning, costs can rise quickly. We need more people with experience in

implementation, and more evaluations of implementations are needed that are not commercially biased.

- Sustainability: All of these systems will require sustainability plans for both funding and training of staff. It is unclear what the maintenance costs will be and who will pay for those costs.
- Data Quality Is Poor: There are poor tools and processes for quality assurance of data, reflected in poor disease registries with incomplete or inaccurate data. Hence aggregate data analysis is unreliable at the regional and national levels.
- People: Most importantly, there are not enough informatics professionals are available with advance training in analytics and evaluation. This is a necessary building block for countries to have sustainable and successful informatics systems and health outcomes.

## CONCLUSIONS

There are many examples of successful global health informatics systems that have resulted in both positive patient outcomes and reduction of costs. However, the literature also shows that we need more and better-designed controlled studies with more participants. We also need clearer documentation of the evaluation methods so we can allow replication of the programs in other regions and make comparisons of implementation costs and outcomes. Major drivers in the development of this field include new technologies, particularly mobile devices and new informatics research methods. Major challenges include the need for data integration, adaptation of systems, and training of staff. We need continued investments in research to understand the effectiveness of the system and to learn how to scale solutions to meet global needs. Measuring outcomes will be a key to providing evidence for the future growth of global health informatics.

## References

Alvarez Flores, M.G., Guarner, J., Terres Speziale, A.M., 1995. [Productivity before and after installing a computerized system in a clinical laboratory]. Rev. Invest. Clin. 47 (1), 29–34.

American Medical Informatics Association. Biomedical Informatics Core Competencies. <https://www.amia.org/biomedical-informatics-core-competencies> (accessed 16.07.16).

Bates, D.W., Leape, L.L., Cullen, D.J., et al., 1998. Effect of computerized physician order entry and a team intervention on prevention of serious medication errors. JAMA 280 (15), 1311–1316.

Biesecker, L.G., Burke, W., Kohane, I., Plon, S.E., Zimmern, R., 2012. Next-generation sequencing in the clinic: are we ready? Nat. Rev. Genet. 2012 13 (11), 818–824.

Blaya, J.A., Shin, S.S., Yagui, M.J., Yale, G., Suarez, C.Z., Asencios, L.L., et al., 2007. AWeb-based laboratory information system to improve quality of care of tuberculosis patients in Peru:

functional requirements, implementation, and usage statistics. BMC Med. Inform. Decis. Mak. 7, 33.

Blaya, J.A., Fraser, H.S., Holt, B., 2010. E-health technologies show promise in developing countries. Health Aff. (Millwood) 29 (2), 244–251.

Canuel, V., Rance, B., Avillach, P., Degoulet, P., Burgun, A., 2014. Translational research platforms integrating clinical and omics data: a review of publicly available solutions. Brief Bioinform. 16, 280–290.

Centers for Medicare & Medicaid Services (2016). Electronic Health Records [Definition]. <https://www.cms.gov/ehealthrecords/>.

Crotty, B.H., Walker, J., Dierks, M., Lipsitz, L., O'Brien, J., Fischer, S., et al., 2015. Information sharing preferences of older patients and their families. JAMA Intern Med 175 (9), 1492–1497, <http://dx.doi.org/10.1001/jamainternmed.2015.2903>. PubMed PMID: 26147401.

Darkins, A., et al., 2008. Care Coordination/Home Telehealth: the systematic implementation of health informatics, home telehealth, and disease management to support the care of veteran patients with chronic conditions. Telemed J. E Health 14 (10), 1118–1126.

Delloitte Center for Health Solutions, 2015. mHealth in an mWorld How mobile technology is transforming health care. <http://www.deloitte.com> (accessed 11.07.16).

Deloitte, 2014. Global health care outlook, Shared Challenges, Shared Opportunities. <https://www2.deloitte.com/content/dam/Deloitte/global/Documents/Life-Sciences-Health-Care/dttl-lshc-2014-global-health-care-sector-report.pdf> (accessed 16.07.16).

Detmer, D.E., Shortliffe, E.H., 2014. Clinical informatics: prospects for a new medical subspecialty. JAMA 311 (20), 2067–2068, <http://dx.doi.org/10.1001/jama.2014.3514>. PubMed PMID: 24823876.

Elbert, N.J., van Os-Medendorp, H., van Renselaar, W., Ekeland, A.G., Hakkaart-vanRoijen, L., Raat, H., et al., 2014. Effectiveness and cost-effectiveness of ehealth interventions in somatic diseases: a systematic review of systematic reviews and meta-analyses. J. Med. Internet. Res. 16 (4), e110.

Evans, D.C., Nichol, W.P., Perlin, J.B., 2006. Effect of the implementation of an enterprise-wide Electronic Health Record on productivity in the Veterans Health Administration. Health Econ. Policy Law 1 (Pt 2), 163–169.

Evans, W.D., Wallace, J.L., Snider, J, 2012. Pilot evaluation of the text4baby mobile health program. BMC Pub. Health 12, 1031, <http://www.biomedcentral.com/1471-2458/12/1031>.

Fraser, H.S., Jazayeri, D., Nevil, P., Karacaoglu, Y., Farmer, P.E., Lyon, E., et al., 2004. An information system and medical record to support HIV treatment in rural Haiti. BMJ (Clinical research ed) 329 (7475), 1142–1146, http://dx.doi.org/10.1136/bmj.329.7475.1142. PMC 527691free to read. PMID 15539669.

Fraser, H.S., Blaya, J., Choi, S.S., Bonilla, C., Jazayeri, D., 2006. Evaluating the impact and costs of deploying an electronic medical record system to support TB treatment in Peru. AMIA Annu. Symp. Proc. 264–268.

Global Health Workforce Alliance, Task Force for Scaling Up Education and Training for Health Workers, 2008. Scaling Up, Saving Lives. May 2008. <http://www.who.int/workforcealliance/documents/Global_Health%20FINAL%20REPORT.pdf>.

Groves, P., Basel, K., Knott, D., Van Kuiken, S.V., 2013. The 'Big Data Revolution in Healthcare. Accelerating Value and Innovation. McKinsey & Company, New York: NY, January 2013.<http://www.mckinsey.com/insights/health_systems_and_services/the_big-data_revolution_in_us_health_care> (accessed 11.07.16).

Hartzler, A., Wetter, T., 2014. Engaging patients through mobile phones: demonstrator services, success factors, and future opportunities in low and middle-income countries. Yearbk. Med. Inform. 9 (1), 182–194.

Hasman, A., Mantas, J., 2013. IMIA accreditation of health informatics programs. Healthc. Inform. Res. 19 (3), 154−161, http://dx.doi.org/10.4258/hir.2013.19.3.154. Epub 2013 Sep 30. Review. PubMed PMID: 24175114; PubMed Central PMCID: PMC3810522. <http://www.ncbi.nlm.nih.gov/pubmed/24175114> (accessed 11.07.16).

Henderson, C., et al., 2014. Cost-effectiveness of telecare for people with social care needs: the Whole Systems Demonstrator cluster randomised trial. Age Ageing 43 (6), 794−800.

HIMSS, 2016. Technology Informatics Guiding Education Reform TIGER. <http://www.himss.org/professionaldevelopment/tiger-initiative> (accessed 11.07.16).

International Journal of Medical Informatics, 2016. About the Journal. <http://www.journals.elsevier.com/international-journal-of-medical-informatics>(accessed 16.07.16).

International Telecommunications Union, 2015. ITU 2015 ICT figures. <http://www.itu.int/net/pressoffice/press_releases/2015/17.aspx> (accessed 11.07.1616).

Jamoom, E., Hing, E., 2015. Progress with electronic health record adoption among emergency and outpatient departments: United States, 2006−2011. NCHS Data Brief (187), 1−8, PubMed PMID: 25714041. <http://www.cdc.gov/nchs/data/databriefs/db187.pdf>.

Jha, A.K., 2011. The promise of electronic records: around the corner or down the road? JAMA 306 (8), 880−881, <http://dx.doi.org/10.1001/jama.2011.1219>. PubMed PMID: 21862751.

Kaushal, R., Shojania, K.G., Bates, D.W., 2003. Effects of computerized physician order entry and clinical decision support systems on medication safety: a systematic review. Arch. Intern. Med. 163 (12), 1409−1416.

Komatsoulis, G.A. "Program Announcement". National Cancer Institute. Archived from the original on July 30, 2012. <http://web.archive.org/web/20120730234757/https://cabig.nci.nih.gov/program_announcement> (accessed 05.07.16).

Koppel, R., 2010. Monitoring and evaluating the use of electronic health records. JAMA 303 (19), 1918, author reply 1918−1919. <http://dx.doi.org/10.1001/jama.2010.590>. PubMed PMID: 20483967.

Kulikowski, C.A., Shortliffe, E.H., Currie, L.M., Elkin, P.L., Hunter, L.E., Johnson, T.R., et al., 2012. AMIA Board white paper: definition of biomedical informatics and specification of core competencies for graduate education in the discipline. J. Am. Med. Inform. Assoc. 19 (6), 931−938.

Kumar, S., Nilsen, W.J., Abernethy, A., Atienza, A., Patrick, K., Pavel, M., et al., 2013. Mobile health technology evaluation: the mHealth evidence workshop. Am. J. Prev. Med. 45 (2), 228−236, <http://dx.doi.org/10.1016/j.amepre.2013.03.017>. PubMed PMID: 23867031; PubMed Central PMCID: PMC3803146.

Kuo, M.C., Chang, P., Feng, R.C., Ball, M., Westra, B., 2012. From passion to excellence: national nursing informatics movement in Taiwan. Comput Inform Nurs 30 (5), 234−236, <http://www.ncbi.nlm.nih.gov/pubmed/22588169>.

Kushniruk, A., Kaipio, J., Nieminen, M., Nøhr, C., Borycki, E., 2013. Comparing approaches to measuring the adoption and usability of electronic health records: lessons learned from Canada, Denmark and Finland. Stud. Health Technol. Inform. 192, 367−371, PubMed PMID: 23920578.

Luna, D., Almerares, A., Mayan III, J.C., González Bernaldo de Quirós, F., Otero, C., 2014. Health Informatics in developing countries: going beyond pilot practices to sustainable implementations: a review of the current challenges. Healthc. Inform. Res. 20 (1), 3−10, <http://dx.doi.org/10.4258/hir.2014.20.1.3>. Epub January 31, 2014. Review.

Mantas, J., Hasman, A., Shortliffe, E.H., 2013. Assessment of the IMIA educational accreditation process. Stud. Health Technol. Inform. 192, 702−706, <http://www.ncbi.nlm.nih.gov/pubmed/23920647>(accessed 11.07.16).

McCaffery, J., 2009. Global Partnerships: Strengthening Human Resources for Health Approach Together, September 2009. <http://www.capacityplus.org/global-partnerships-strengthening-human-resources-health-approaches-together> (accessed 16.07.16).

McConnell, P., Dash, R.C., Chilukuri, R., et al., 2008. The cancer translational research informatics platform. BMC Med. Inform. Decis. Mak. 2008 8, 60.

McCormick, K.A., Delaney, C.J., Brennan, P.F., Efren, J.A., Kendrick, K., Murphy, J., et al., 2007. Guideposts to the future—an agenda for nursing informatics. J. Am. Med. Inform. Assoc. 14 (1), 19–24, <http://www.ncbi.nlm.nih.gov/pmc/articles/PMC2215078/> (accessed 11.07.16).

Mihalas, G., Zvarova, J., Kulikowski, C., Ball, M., van Bemmel, J., Hasman, A., et al., 2014. History of medical informatics in Europe—a short review by different approach. Act. Inform. Med. 22 (1), 6–10, <http://dx.doi.org/10.5455/aim.2014.22.6-10. Epub 2014 Jan 25. Review. PubMed PMID: 24648613; PubMed Central PMCID: PMC3947944. <http://www.ncbi.nlm.nih.gov/pubmed/24648613> (accessed 11.07.16).

Mobile World Center, 2016. <http://www.mobilehealthglobal.com> (accessed 11.07.16).

Mohla C., 2014. Health IT Workforce Training Programs: Looking Back, Looking Forward. <http://www.healthit.gov/buzz-blog/community-college-consortia/health-it-workforce-training-programs-reflection/http://www.healthit.gov/buzz-blog/community-college-consortia/health-it-workforce-training-programs-reflection/> (accessed 11.07.16).

Murphy, S.N., Mendis, M., Hackett, K., Kuttan, R., Pan, W., Phillips, L.C., et al., 2007. HC: Architecture of the open source clinical research chart from Informatics for Integrating Biology and the Bedside. AMIA Annu. Symp. Proc. 2007, 548–552.

National Cancer Institute, 2015a. caBIG®—Cancer Biomedical Informatics Grid®. <https://cabig-kc.nci.nih.gov/> (accessed 05.07.16).

National Cancer Institute, 2015b. Center for Biomedical Informatics and Information Technology (CBIIT). caGrid®. <http://cagrid.org> (accessed 11.07.16).

National Cancer Institute, 2015c. i2b2: Informatics for Integrating Biology and the Bedside (2015). <http://www.i2b2.org> (accessed 11.07.16).

National Institutes of Health, 2011. An Assessment of the impact of the NCI Cancer Biomedical Informatics Grid (caBIG®). Report of the Board of Scientific Advisors Ad Hoc Working Group, Cancer Biomedical Informatics Grid (caBIG®) Program National Cancer Institute. March 2011.

National Library of Medicine, 2016. Health Informatics. Original from Procter, R. Dr. (Editor, Health Informatics Journal, Edinburgh, United Kingdom). Definition of health informatics [Internet]. Message to: Virginia Van Horne (Content Manager, HSR Information Central, Bethesda, MD). August 16, 2009 (cited September 21, 2009) <https://www.nlm.nih.gov/hsrinfo/informatics.html> (accessed July 16, 2016).

NORC at the University of Chicago, 2014. FINAL REPORT Evaluation of the Information Technology Professionals in Health Care ("Workforce") Program—Summative Report—March 2014. <http://www.healthit.gov/sites/default/files/workforceevaluationsummativereport.pdf> (accessed 11.07.16).

Pearl, R., 2014. Kaiser Permanente Northern California: current experiences with Internet, mobile, and video technologies. Health Aff. (Millwood) 33 (2), 251–257, <http://dx.doi.org/10.1377/hlthaff.2013.1005>. Review. PubMed PMID: 24493768.

Pew Research Center, 2015. U.S. Smartphone Use in 2015 Pew Research Center. <http://www.pewinternet.org/2015/04/01/us-smartphone-use-in-2015> (accessed 11.07.16).

Quintana, Y., Patel, A.N., Arreola, M., Antillon, F.G., Ribeiro, R.C., Howard, S.C., 2013. POND4Kids: a global web-based database for pediatric hematology and oncology outcome evaluation and collaboration. Stud. Health Technol. Inform. 2013 183, 251–256.

Reardon, S., 2011. A world of chronic disease. Science 333 (6042), 558—559, <http://dx.doi.org/10.1126/science.333.6042.558>.

Robertson, J., Walkom, E., Pearson, S.A., Hains, I., Williamsone, M., Newby, D., 2010. The impact of pharmacy computerised clinical decision support on prescribing, clinical and patient outcomes: a systematic review of the literature. Int. J. Pharm. Pract. 18 (2), 69—87, Review. PubMed PMID: 20441116.

Rotich, J.K., Hannan, T.J., Smith, F.E., Bii, J., Odero, W.W., Vu, N., et al., 2003. Installing and implementing a computer-based patient record system in sub-Saharan Africa: the Mosoriot Medical Record System. J. Am. Med. Inform. Assoc. 10 (4), 295—303, Epub 2003 Mar 28.

Safran, C., 2009. Informatics training for clinicians is more important than hardware and software. Yearbk. Med. Inform 164—165.

Safran, C., Rind, D.M., Davis, R.M., Currier, J., Ives, D., Sands, D.Z., et al., 1993. An electronic medical record that helps care for patients with HIV infection. Proc. Annu. Symp. Comput. Appl. Med. Care 224—228, PubMed PMID: 8130466; PubMed Central PMCID: PMC2248507.

Saltz, J., Oster, S., Hastings, S., et al., 2006. CaGrid: design and implementation of the core architecture of the cancer biomedical informatics grid. Bioinformatics 2006 22, 1910e6.

Seebregts, C.J., Mamlin, B.W., Biondich, P.G., Fraser, H.S., Wolfe, B.A., Jazayeri, D., et al., 2009. The OpenMRS Implementers Network. Int. J. Med. Inform. 78 (11), 711—720.

Sequist, T.D., Cullen, T., Hays, H., Taualii, M.M., Simon, S.R., Bates, D.W., 2007. Implementation and use of an electronic health record within the Indian Health Service. J. Am. Med. Inform. Assoc. 14 (2), 191—197.

Sipes, C., 2016. Project management: essential skill of nurse informaticists. Stud. Health Technol. Inform. 225, 252—256, PubMed PMID: 27332201.

Smart, D.R., 2012. Physician Characteristics and Distribution in the US. 2012 Edition American Medical Association Press, Chicago, IL.

Stevento, A., et al., 2012. Whole System Demonstrator Evaluation Team. Effect of telehealth on use of secondary care and mortality: findings from the Whole System Demonstrator cluster randomised trial. BMJ 344, e3874.

United Nations, Department of Economic and Social Affairs, Population Division, 2015. World Population Ageing 2015 (ST/ESA/SER.A/390). <http://www.un.org/en/development/desa/population/publications/pdf/ageing/WPA2015_Report.pdf> (accessed 16.07.16)

World Health Organization, 2005. 58th World Health Assembly Report; Geneva: WHO; 2005. <http://www.who.int/healthacademy/media/WHA58-28-en.pdf> (accessed 16.07.16).

World Health Organization, 2016. Global Observatory for eHealth. <http://www.who.int/goe> (accessed 11.07.16).

# CHAPTER 2

# National Health Information Systems and Health Observatories

**R.J. Rodrigues[1,2] and C.H. Gattini[3]**
[1]eHealthStrategies, Bethesda, Maryland, United States
[2]Global Health International Advisors, Washington DC, United States
[3]University of Chile, Santiago, Chile

## CONTENTS

## NATIONAL HEALTH INFORMATION SYSTEMS AND HEALTH OBSERVATORIES: RESOURCES FOR EVIDENCE-BASED DECISION-MAKING

Aiming to achieve better services and outcomes, national health authorities develop strategies and implement plans directing and sustaining the improvement of the sector infrastructure and interventions. Decisions and ensuing actions must be supported by evidence to guide resource allocation and management, performance measurement, the routine operation of a large number of technical areas, and administrative actions at different levels of the health system.

Health information systems are sets of organized and interacting components around the goal of producing information aimed at supporting those decision-making and actions at each level of a health system and healthcare organization. Health information is a key input for health-related policy development, program planning and operation, health surveillance, and outcomes monitoring, evaluation and research.

National Health Information Systems (NHIS) and Health Observatories (HOs) are two of those informational resources. They are essential components of Global Health—a comprehensive interdisciplinary area of knowledge, policy, and practice that seeks to improve individual and collective health, achieve health equity, and contribute to worldwide socioeconomic development through transnational partnerships and intersectoral collaboration.

**Global Health Informatics.**

- NHIS and HO involve resources, technology, and processes that have a key role in the continuous monitoring and evaluation of progress toward the achievement of the proposed health and healthcare operational and outcome targets.
- The conceptual emphasis of a particular information system may be different, according to the prevailing institutional perspective and goals. The focus may be on data, processes, technology, institutional organization, contents, or particular aspects of health system and healthcare activities.
- The operational nature of health information systems also varies among different organizations. They may support only specific technical or administrative functions, have a broad institutional transdepartmental role, or be a mixture of standalone and connected applications. Yet, in all cases, they are linked to one or more core operational components of the health institution.
- NHIS and HOs are key resources that support decision-making and action by (1) providing "information for action"; (2) showing results and providing evidence for future interventions; and (3) upholding accountability.

Health organizations and agencies increasingly depend and compete on the quality of their analytics and their ability to make correct decisions. Having aligned objectives, incentives, and performance measurements, organization-wide analytic systems are essential for coordinated action.

- There is a need for health professionals and teams to look at the same key indicators to uphold a single view of performance throughout the organization.
- Gaps in underlying data, poorly designed data queries, inadequate visualization, lack of standardization, and poor client experiences are common reasons for underutilization of analytics in the generation of knowledge and intelligence resulting in poor end-user engagement.

## National Health Information System

A NHIS is the organized and integrated network of resources and processes that contribute to the overall production and communication of nationwide health-related information. Relevant characteristics are:

- The ultimate aim is to assist policymakers and healthcare practitioners in preserving and improving the health of the population.
- NHIS is defined and coordinated by national health authorities, with the primary aim of supporting evidence-based decision, action, and documentation of health status indicators.

- NHIS is composed of several routine and nonroutine specialized information applications maintained by various programs, offices, and institutions operating and interacting at different levels of the country health system. It includes paper-based documentation and electronic databases.
- Depending on the country health systems model and its level of integration, diverse organizational patterns of health information systems exist, ranging from centrally planned and integrated systems to heterogeneous and disconnected arrangements characterized by redundant, fragmented, and limited data.
- Overall performance and production of a NHIS depend on the degree of data reliability, efficiency, coordination, and integration of all the specific health information subsystems of the nationwide information network.
- Ideally, the data capture, processing, and generation of information collected from service utilization and production statistical sources and from medical record systems should be uniformly defined and consistently applied to all areas and levels of the national health system.
- Specific information subsystems vary in nature and setup, depending on the institutional formal or functional organization, degree of automation, as well as the complexity of the institutional area to which they belong.
- The informational content addresses (1) health situation in terms of morbidity and mortality; (2) health-related risks and determinants; (3) health systems' resources (financial, human, physical, and technological); (4) processes; (5) performance in the provision of healthcare; and (6) the outcomes of health interventions.
- The NHIS, depending on the level of development of the health sector in the country, may include a wide range of key population-oriented basic data and indicators, as well as textual, qualitative, and graphic information. They take account of (1) the government steering role (regulation, policies and plans, advocacy, research dissemination); (2) quantitative and qualitative data on resources (administrative structure, budget, costs, employees, direct care professional profile, physical infrastructure, medical diagnostic, and therapeutic technology); (3) health services delivery (health promotion; preventive-oriented services, curative-oriented services, rehabilitative services, and palliative services), and (4) environmental action (programs and their control).
- The NHIS is responsible for the production of reliable and timely information, but is not accountable for its appropriate use. Beyond the quality and potential value of the produced health information, its value depends on its actual use by policy- and decision-makers, as well as by other users who need evidence to orient interventions.

- Less-developed countries tend to have limited and inefficient health systems and inadequate routine health information systems. Aiming to improve global health by strengthening health information, in 2005 the World Health Organization launched the Health Metrics Network (HMN)—a global partnership dedicated to the strengthening of health information systems in low and low-middle-income countries.

## Health Observatory

A HO is a policy-oriented national center designed to perform ongoing and systematic observation and analysis of relevant health issues related to population groups and healthcare organizations. HOs are "intelligence-generating units" with the goal of supporting timely and effective evidence-based decision-making and action in public health and for the operation of health systems.

One distinguishing character of HOs is that they are deployed in a virtual environment, in the sense of being a construct that exists as an electronic public network application as opposed to a traditional repository of physical paper documents. This feature of HOs' character emphasizes that they are not an entity based in physical documentation (libraries, printed reports, historical record collections, patient medical records, etc.) but are a dynamic digital resource that take advantage of databases and analytical and display software, providing access to information, knowledge, and intelligence in a variety of standardized and user-defined formats.

- To observe is the process of paying close attention to someone or something that is happening or might take place, using instruments and tools to properly detect and record an observed situation in a defined context.
- Observation leads to "overviews," i.e., the construction of a broad appraisal, a general or comprehensive outline or summary of a defined health situation under interest—including health status and health risks and determinants—related to the population of a country or defined geographical area for a defined period or at an explicit point in time.
- Perform systematic, continuing, and integrated health-related observations, producing regular and ad hoc reports with updated overviews focused on selected or emergent issues relevant to key policy and decision-making. A HO is an operational- and problem-solving-oriented tool, with intrinsic capacity for scientifically oriented analysis and research.
- The ultimate goal is to contribute to the preservation and improvement of health of the population, including the reduction of inequalities, and serve as a tool able to provide valid and comparable overview or

scenarios for health planning including cross-sectional variations with implications to equity and alternative scenarios.

■ Main users are health policymakers, planners, and managers at all levels of the health system as well as political and health authorities.

Using creative and evidence-based approaches, it operates as a "health intelligence unit" with capacity for collecting, analyzing, reporting, and sharing information, thus generating and adding new knowledge to the field. As intelligence generation centers, HOs take advantage of data already collected and managed by the NHIS and, as required, data captured by program-specific subsystems such as surveillance and auditing.

■ Provide policy-oriented information and intelligence, statistical results, contextual analyses, and forecasting as elements of routine reporting or answering specific requests from high-level policymakers, planners, and managers aimed at the improvement of national capacity including the provision of guidelines, methods, techniques, and training.

■ As a resource for evidence-based decision-making, the observatory produces dynamic and effective information, knowledge, and intelligence for assessment, analysis, and action based on trends, as well as current and projected scenarios, in support of planning and decision-making.

■ Perform health surveillance either by integrating contents or functions of a public health surveillance system or by executing that function by default.

■ Carry out the monitoring of health systems, either by integrating contents and functions of a monitoring system or by performing that function by default. This includes the publication of comparable national and subnational profiles on health status and health system components.

■ Support knowledge management and sharing customized to primary audiences (policy- and decision-makers) and, according to priorities and possibilities, to secondary users.

## DATA, INFORMATION, KNOWLEDGE, AND INTELLIGENCE

Data, information, knowledge, and intelligence make up the intellectual capital of an organization. The generation of knowledge and intelligence requires a collaborative and integrated approach to the capture, organization, access, and use of the pertinent institutional, sectoral, and partner data to support the backsight and foresight needed to respond appropriately to a variety of internal and external decision-makers and other civil society users.

## From Data to Intelligence: Progressing From "Know That" to "Know How"

Data are factual descriptive items of reality. In the case of NHIS and HOs, data items deemed of importance to specific human health and healthcare domains are collected in standardized formats and assembled into predefined datasets. Birth and mortality, morbidity profiles, patient services and resource utilization, service delivery costs, determinants of health, disease surveillance, and intervention outcomes are common examples of those sets.

Professional and scientific bodies, national regulatory offices, international agencies, and policymakers are responsible for the definition of the datasets employed in each case. Those definitions may differ and change over time according to perceived needs, local considerations, scientific interest, or national and international standards.

### Information Production

Collected health data of different types and sources are collated, ordered, and compared with the goal of establishing linkages among data items to generate information, i.e., the construction of statements about the reality that have meaning to the user by being significant to activities or decisions related to a specific health domain.

- Information is data with meaning. It provides qualitative and quantitative answers to posed question. Information conveys different meanings in different contexts.
- Information is produced by operations realized upon a specific domain dataset according to preestablished rules and carried out manually, mediated through technological resources, or by a combination of both.
- The information about a health domain is nearly always represented on a timescale.
- Information is the output of health information systems and a key input for the acquisition of knowledge, better policymaking, management, selection of possible actions, and performance monitoring.
- Data and information can be conceptualized by the statement "know that."

### Knowledge as Information in Context

Knowledge differs from information by involving judgmental elements and expertise or understanding of the natural world that blends facts, information, descriptions, or skills acquired through experience or education.

Knowledge acquisition involves complex cognitive processes—perception, communication, and reasoning.

- The generation of knowledge, differently than information production, always involves human technical decisions and judgment.
- Knowledge is information in context involving the merger of information from different sources and frequently from different domains.
- Knowledge generation and dissemination is a process whereby internal and external knowledge circulates and consolidates the functional relationship and common understanding of the reality among different organizations or units of a single organization.
- Knowledge management plays a key role in sustainable organizational change.

### Intelligence as Knowledge Translation

Higher levels of conceptualization of reality can be further construed when an experienced individual analyzes knowledge acquired from different perspectives or thematic areas and compares them with the published experiences of other professional authorities. This stepwise aggregation of knowledge results in intelligence, a human activity requiring access to knowledge bases, specialized expertise, and highly developed analytical capability. Experts, information, knowledge, and intelligence form an integral unit—each empowers the others.

The generation of intelligence from informational sources involves incorporating knowledge from other sources than primary domain references. Particularly when dealing with clinical knowledge and intelligence, informational sources other than those originating from healthcare primary datasets and include the results of randomized controlled trials; systematic reviews or metaanalyses; and third-generation knowledge such as practice guidelines, decision aids, or care pathways based on best available evidence.

The term knowledge translation (KT) appeared more recently in the health literature and chiefly pertains to the broad assessment, review, and utilization of scientific research to answer specific questions. The Canadian Institutes for Health Research (CIHR) defines KT as the exchange, synthesis, and ethically sound application of knowledge within a complex set of interactions among researchers and knowledge users to accelerate the capture of the benefits of research through improved health, services that are more effective, and an efficient healthcare system.

- From our perspective, the term intelligence is interchangeable with the concept of KT.

- Intelligence aims at closing the knowledge-to-action gap in healthcare processes by reducing barriers and taking advantage of facilitators.
- Intelligence is a more extensive and organized form of knowledge supporting the construction of evidence-based statements and of a higher value from the point of view of trust and confidence that it truly represents reality.

To be an effective decision-making tool, information, knowledge, and intelligence must be:

- responsive to specific needs for evidence-based decision-making in an opportune and effective way. They must be able to accommodate specific routine and nonroutine requests for information from different managers at different levels and responsibilities.
- coherent with the institutional and managerial culture. They must supply content that reflects evidence and, hopefully, be able to produce the type of analysis required by decision-makers.
- provide evidence reflecting—at least partly and indirectly—the situation that it is assumed to represent (health, health needs, health systems, and services) following agreed and clear standards and assumptions on each data and indicator.
- exhibit quality, completeness, opportunity, and validity and a balance between definitive official (historical) information and draft (provisory) initially produced information.
- knowledge and intelligence can be conceptualized by the statement "know how."

## EVIDENCE-BASED PRACTICE

Since the early1990s, the use of best current evidence has been gaining impetus in decision-making in health services program management and clinical practice. Evidence-based health practice (EBHP) replaces the traditional model of "medicine by authority" and the use of unendorsed and nonsystematic information. EBHP is a process of systematically locating, appraising, and using validated knowledge and intelligence sources, research, and critical reviews.

- Possibly the greatest contribution of informatics to health practice is the support provided by digital and communication technologies to the information- and knowledge-intensive process needed for EBHP.
- NHIS and HOs are only one of the many sources of data, information, knowledge, and intelligence that support EBHP (Fig. 2.1).
- NHIS and HOs maintain data, information, knowledge, and intelligence acquired from contextual environmental, epidemiological,

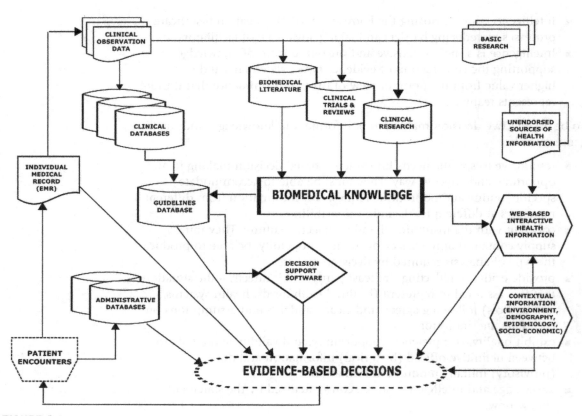

**FIGURE 2.1**

Environmental, socioeconomic, demographic, and epidemiological contextual data residing in NHIS and HO databases is one of the many components of the universe of health data sources needed for EBHP. *NHIS*, National Health Information Systems; *HOs*, Health Observatories; *EBHP*, evidence-based health practice.

socioeconomic, and case-specific data from repositories of medical and diagnostic encounters.

- Environmental, socioeconomic, demographic, and epidemiological contextual evidence is generally short-lived, limited in geographic scope, and not subject to the more rigorous systematic review process followed by formal literature publications. The central issue in those categories of evidence is the specificity, time-sensitive, and the opportune collection and availability of data for analysis.

- Environmental, socioeconomic, demographic, and epidemiological data are only valid to a specific site and timeframe, but they carry high informational value for decision-makers. As example, the weekly reporting on the incidence of communicable diseases illustrates the importance of contextual information as evidence in support of clinical diagnosis and identification of emerging diseases.

- Information systems and, particularly, Internet-based interactive health communication technologies and expert systems, offer great potential in (1) accessing evidence-based knowledge on demand; (2) the possibility of creating interactive objects, tools, and resources; and (3) supplying an enhanced ability to provide widespread dissemination and immediate updating of content and functionalities of clearinghouses aimed at corporate and public information.
- Data warehousing applications permit digital data from various sources to be merged and accessed by any authorized user and, when used in conjunction with knowledge engineering techniques and products, support the generation of information, knowledge, and intelligence to timely respond to new or emerging demand for information coming from changes in information needs required by a dynamic management.
- Digitized clinical, administrative, and financial data can be mined and exploited to support outcome-based clinical processes, research, surveillance, and health interventions.
- From a population perspective, collections of structured records can be maintained in clinical and administrative repositories with data that can be selectively searched and extracted.

## HISTORICAL, SURVEILLANCE, AND MONITORING DATA

Historical, surveillance, and monitoring systems provide the background information, knowledge, and intelligence to understand, manage, and update the health sector and promote better outcomes.

- To measure the impact of interventions or changes in demographics, it is essential to develop national or regional chronological health profiles, particularly of morbidity, mortality, and service utilization coupled with the deployment of appropriate resources for the evaluation of broad changes of those profiles over time.
- Standard sets of data and indicators are collected on a regular basis to produce official statistical reports. Historical series are constructed and reports summarize the values, variations, trends, evaluation, and resulting information about the population of a country or political unit. Most national health systems follow this pattern by adopting datasets recommended by international organizations.
- Historical series make the main building blocks of a NHIS. Yet, historical information is usually highly aggregated and thus of limited value to day-to-day health practice and managerial decision-making.
- According to the US Center for Disease Control and the World Health Organization, health surveillance is the ongoing systematic collection, analysis, and interpretation of specific health datasets essential for

planning, implementing, and evaluating public health activities. It is closely integrated with timely analysis and dissemination of the relevant data to enable effective and efficient action to prevent and control disease.

- Surveillance systems track a defined health situation and provide alert for opportune and effective action, if necessary. Surveillance systems focus is a defined target population (population approach).
- Surveillance systems are important in public health particularly regarding agent or hazard, risk factor, exposure, or health event and essential for decisions regarding interventions and warnings or cautionary information for a defined geographical area and target population at a defined time.
- Monitoring is the acquisition of systematic knowledge regarding critical components of a program, project, or strategy. The goal of monitoring is to establish the extent by which the delivery of goods and services are occurring in accordance with what was planned. Monitoring is essential in guiding what appropriate measures are necessary to correct detected deficiencies.
- Monitoring health systems involves gathering ongoing core information on health systems' structure and performance, as well as their target populations, then analyzing, interpreting, and timely reporting the results to those who are responsible for health systems and services-related policy and action.
- Monitoring systems have an action-oriented approach, similar to the approach and methods followed by surveillance systems. Both track-specific conditions over time, comparing the level of an "observed" situation with an "expected" situation to detect gaps, alert signals, or needs for corrective action.
- Health-related surveillance and monitoring are the two main methodological approaches featured in HOs. Surveillance is mainly focused on population health while monitoring spotlights health systems performance. Although there may be differences in specific situations being studied, all share similarities in the approaches and methods used to observe health-related issues.

## HOS HAVE SPECIFIC NEEDS FOR DATA AND DATA MANAGEMENT

Although the large variety of health domains results in extensive demand for health intelligence, HOs, by nature, have a focused scope to a few domains. In their design, one must establish priorities regarding data capture, information processing, and the demanding analytical work needed for the production of actionable intelligence. Not necessarily being repositories of large amounts of data, a HO uses extant data from many sources but frequently also targets very specific areas for data collection.

- Many HOs are the product of the particular circumstances that led to their establishment, distinguishing them from NHIS that typically collect well-established datasets on a continuous basis over a long time span and designed at the historical follow-up of basic health indicators.
- The "one of its kind" or problem-oriented HO have their own particular goals and datasets. They may be relatively short-lived and only operate while the specific health problem exists.
- Public Health Observatories (PHOs) strive to combine the qualities of academic research and the practice of public health institutions by providing on a shorter timescale reliable, timely, and high-quality health intelligence of higher granularity regarding geography, to enable rapid responses focused at current health issues at the local, regional, urban, and neighborhood level.
- Resiliency and fast response in developing and implementing new data collection and processing routines require strong networks of professionals that must be comparatively autonomous and capable of developing their own datasets and intelligence-generating operations to answer local needs. Those needs reflect the increasing value placed on cross agency work, the identification of health inequalities, and the importance of evidence-based policymaking.

## FUNCTIONAL COMPONENTS

The functional components are common to both NHIS and HO and fit into three categories: (1) data sources, data domains, and data management and operations (inputs); (2) production of indicators, knowledge, and intelligence; and (3) communication of products (outputs) (see Fig. 2.2).

### Data Sources, Data Domains, and Data Management and Operations

*Data Sources*—internal and external sources of raw and treated data and indicators for the extraction of selected sets of data elements, identified as required for corporate and external information and for the production of knowledge and intelligence.

*Data Domains and Data Elements*—types of data and their descriptors, metadata, standards for capture, quality control, and entry procedures, as required for corporate and external information and production of knowledge and

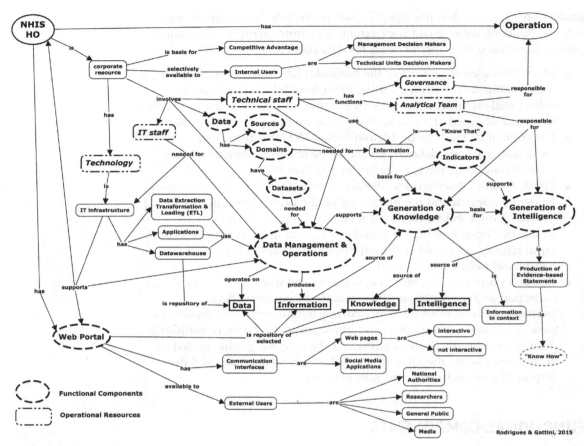

**FIGURE 2.2**

Conceptual map of NHIS and HO. *NHIS*, National Health Information Systems; *HOs*, Health Observatories.

intelligence. Major challenges to system designers relate to issues of awareness of needed data elements ("need to know" data) and its existence and availability or feasibility of discovery and capture.

*Data Management and Operations*—information technology resources for the extraction, transformation, loading, warehousing, access, manipulation and display. They include a business intelligence (BI) accessibility layer capable of (1) generating information; (2) processing sets of indicators; (3) providing support to data analysis; (4) producing standardized and ad hoc reports; and (5) interfacing with visualization tools.

- Standardization is a necessary prerequisite for the upstream consolidation of case and administrative data and record integration.

It requires unique identifiers, uniform clinical encounter datasets, and classification terminologies, in most cases nonexistent in the current health system.

■ Lack of data structure, terminology incompatibilities, the separation of clinical from financial and administrative data, and episode fragmentation—the breakup of patient data over time and geographic space, all residing in different physical sites under different patient identifiers—disallows a systematic utilization of the wealth of data residing in manual nonautomated medical and administrative records.

## Data Analysis and Production of Indicators, Knowledge, and Intelligence

The objective of data analysis is to generate evidence-based statements. The generation of knowledge and intelligence is a human activity requiring specialized skills that no single individual detains. The generation of knowledge and translation of knowledge into evidence-based statements (intelligence) involves personal choices, lessons learned, and a "philosophical stance", i.e., a body of beliefs and a thoughtfully constructed logical argument used to explain a viewpoint regarding (1) the nature of the event about which a statement is made; (2) the particular environment to which the statement refers; and (3) the significance of social determinants, where ethical, legal, and political considerations play an important role.

■ Involves proficiency in the use of digital systems and a variety of data operation and application software combined with expertise in one or more health areas.

■ The expertise is required in a permanent basis—and supplemented as required by other internal and external professionals—includes fields such as epidemiology, statistics, information systems, health systems management, and healthcare practice.

## Communication of Information, Knowledge, and Intelligence

Visibility of knowledge and intelligence is a major problem and most organizations do not have good real-time and consistent visibility of their processes. This lack of visibility is not surprising, considering the fact that health systems have complex business rules that, in addition, may be inconsistent across organizational technical areas or may include unique exceptions.

■ The difficulties found in communicating meaningful analyses of data are a direct reflection of the complexity of health systems and organizations.

- Existing information technology infrastructure, software, databases, and applications of an organization, usually are the result of particular development histories and several iterations that occurred over years, representing a mixture of legacy and more current systems—none of which may be appropriate for current purposes.
- The most effective and efficient communication mode is a Web Portal—a web-based repository of data, information, health facts, and results of the analyses and knowledge. Communication with users is accomplished through interactive and noninteractive web pages and social media applications using public networks providing predefined data tables, health facts, and results of the analyses with the objective of making knowledge and intelligence easily accessible.
- Visibility of knowledge and intelligence is enhanced by the use of graphic data display, such as dashboards, and metrics applications.
- Dashboards have become popular in recent years as uniquely powerful tools for communicating important information at a glance. Dashboard implementations take advantage of a variety of graphical displays.
- A dashboard might be the visible outcome of metrics providing a high-level view of relevant results in a single reporting interface. Dashboards often include information on whether any given metric is within its desired range.
- Although dashboards are potentially powerful, this potential is rarely realized—the greatest display technology will not solve the issue of limited knowledge and intelligence visibility without concomitant successful visual design.
- The production of content for the Web Portal requires writing, electronic, and social media communication skills that no single individual masters.

## OPERATIONAL RESOURCES

NHIS and HO operational resources include governance, technical staff, information technology professionals, and information and communication technologies (see Fig. 2.2).

### Governance

In order for information, knowledge, and intelligence to achieve a deliberately engineered change in each setting, it is necessary to (1) assess barriers and facilitators to their generation and use; (2) select, tailor, and implement interventions; and (3) define monitoring parameters. Important issues to be addressed are:

- The technical unit responsible for the data-related and analytical tasks of NHIS and HOs and their professional staff ideally should be under the direct control of the higher administrative level. That location (1) facilitates the coordination of technical unit information systems and corporate databases; (2) encourages and put into effect the standardization of software and data-related procedures; (3) strengthens dealings with the external sources and national authorities; and (4) provides the line authority needed to enforce tasks on a corporate basis.

- Identification of the knowledge-to-action gaps and needs for action, regarding (1) purpose of the system; (2) context (national, regional, and local); (3) perspective (population, organization, provider, health problem, economic issues); (4) datasets required; (5) operational procedures; and (6) resources.
- Closing the consensus gaps among experts by qualitative approaches such as focus groups, interviews, and questionnaires; analysis of existing data by regressing potential determinants of variation in selected data; and analysis of studies concerning guideline implementation.
- Establishment of a managerial and regulatory function and establishment of networks of information flows and processes between various entities with defined and standardized information-related elements—functional components and operational resources.
- The consideration of sustainability should occur early in the process and include the discussion of human, budgetary, managerial, and technological resources.

## Technical Staff

Managing the issues of speed, complexity, and volume associated to analytical work cannot be addressed only in terms of hardware or software—qualified expertise in health practice and health systems administration is essential. This fact has been repeatedly demonstrated in practice and is consistent with the published experience in commercial and industrial applications, where around 75% of the time of professionals is spent in data manipulation activities, even when tools are available for interactive analysis and for creating visualizations aimed at the delivery of an enhanced insight into data.

- Issues that hamper optimal generation of intelligence in health organizations include (1) inadequate technical domain expertise; (2) lack of analytical expertise; (3) limited technological resources; (4) poor communication and understanding between information technology professionals and healthcare staff; (5) communities of experts working with conflicting data definitions; (6) different terms and vocabularies; and (7) inability to search or access pertinent data.

- The analytical work needed for the generation of knowledge and intelligence—including data preparation and the writing of reports—is by far the most demanding and time-consuming activity, frequently suffering from (1) limitations of human resources for conduction of analytical work; (2) existence of departmental information silos; and (3) lack of continuity in collaboration among different internal and external technical units.
- Besides the hard data maintained by NHIS and HOs, frequently there is a rich source of accumulated explicit and implicit experiential knowledge in the institutions and external partner networks—truly a "collective memory" that must be tapped for analytical purposes.
- The technical unit responsible for the data-related and analytical tasks is charged with the following functions:
  - Data-related tasks
    - Operate information technology resources for data extraction, transformation, loading, warehousing, access, manipulation and display including the BI accessibility layer sitting on top of a data warehouse for the generation of reports and analyzes.
    - Build upon existing health information activities of technical units strengthening data compilation and analysis on mandated priority health topics.
    - Promote use of international standards for (1) defining, storing, and managing data terminologies; (2) metadata; indicator definition; (3) data exchange standards; and (4) analytical methods including estimates and adjustments which are used to produce information.
    - Participate and propose the identification of datasets required for corporate and external information, decision-making, and competitive advantage regarding other organizations including guidance in the establishment of internationally adopted databases.
    - Coordinate procedures for extraction of datasets from technical unit databases, enhance data quality, and improve comparability by supporting technical units to address data gaps in terms of availability and consistency.
    - Coordinate with technical units regarding data confidentiality and security issues.
  - Knowledge and intelligence sharing tasks
    - Liaise with partners, including international agencies, private sector, and academic institutions for access to external databases.
    - Define user levels of access to datasets maintained in the data warehouse and support corporate users in the use of analytical interface of the data warehouse.

- Conduct detailed analytical work utilizing internal and external resources when specific technical questions arise.
- Respond to queries from internal and external professionals (government, academic, research, media, and general public)
- Communication of results of analytical studies through a Web standard dissemination platform (Web Portal) and publications;

## Information Technology Professionals

Experience has shown that information technology professionals (1) often do not understand the complexity of the health sector; (2) have a too optimistic and ambitious view for health and healthcare information systems; (3) are not aware of the vagaries of health data definitions and problems of interpersonal relationships among health practitioners, scientists, and health managers; and (4) play down the lack of data processing and analytic skills of many health professionals.

Furthermore, computer professionals, in their desire to perform according to planned project definitions and timetable, have a tendency to hold fast to inadequate early decisions and resist changes to the inadequate initial systems specification which, however, become evident as the health professionals eventually acquire a better grasp of needs, potentialities, and limitations of the proposed health information application in development.

## Information and Communication Technologies

Technology distribution and access deficiencies are common issues in the deployment of national information systems and HOs. In a more limited focus, the "digital divide" encapsulates the dramatic worldwide variation in access to computer-based information technologies, typically measured in terms of teleaccessibility, computer ownership, and Internet connectivity available to individuals and communities.

## SYSTEMS ARCHITECTURE

A typical systems architecture implementation has:

- *An integration server*—hosting all data integration processes (ETL processes) and software applications assigned to (1) the extraction, transformation, and loading of defined datasets from program or theme-specific databases; (2) transformation of extracted data into a standard representation; and (3) uploading them to the data warehouse component. This server may link to a project-specific variable number of servers and services.

- *Database servers*—hosting program-specific databases (source databases), the data warehouse, and the application database. Those servers host all data sources required for analysis, such as transactional databases, datasets, and the data tables from program-specific systems and from external sources. Here users will have access to metadata to explore (1) data element name and code; (2) data type and format; (3) data item definition; (4) technical notes; (5) means of creation, purpose, time and date of creation, and creator or author of the data; and (6) location where the data were created on a computer network. An item of metadata may describe not only an individual datum but also a collection of data including multiple content items and their corresponding hierarchical levels.
- *Query and display facilities server*—with the "BI" core, as an accessibility layer sitting on the top of a data warehouse to make the underlying data more useful through better reporting and improved support for data analysis. It is supplied with a set of tools that allows users to access the data warehouse in a structured manner and includes tools for data querying and dynamic interaction.
- *Data exchange services*—data exchange capacity with external databases from partner institutions for the automatic data extraction and uploading data to the data warehouse. Implementation of Application Programming Interfaces (API) specifies how software components of external and internal systems should interact with each other including libraries with specifications for routines, data structures, object classes, and variables.
- *Web portal*—repository of data, information, health facts, virtual documents, and the compiled results of the analyses and knowledge related to the area of public health presented through interactive and noninteractive web pages and social media applications.
- *Client workstations and remote users*—connected by Local Area Network or a public network.

## Data Processing

Modern data processing systems use data warehouses—long-term, comprehensive data stores in which data items are represented in a denormalized schema that is intended to be more general than the schema of transactional databases. The common data model of the data warehouse is designed to facilitate analysis and uncomplicated to exploit by nontechnical users.

- A data warehouse maintains integrated consistent datasets by extracting selected program-specific data elements residing in a standalone highly detailed program or domain-specific transactional databases.

- The data warehouse standardized datasets support strategic and operational activities and the generation of statistics, knowledge, and evidence-based statements (intelligence/KT).
- Data is imported into the data warehouse from different internal and external transactional databases using ETL (extraction, transformation, and load) processes.
- Metrics applications address the variables to be tracked and define the design and development of the infrastructure required to collect and validate data and make datasets accessible to other modules of the processing setup.
- By implementing international standards for metadata and semantic content, definitions for indicators are constructed in order to be shared and implemented in multiple computer systems. Complete data dictionaries must be prepared—a labor-intensive and frequently neglected task that is required for data comparability.
- Common problem relate to query interfaces not designed for analytic work requiring a level of computer expertise uncommon among health professionals.
- Since data warehouses are enterprise-wide, they respond slowly to changes in a single business unit—in particular, changes introduced into any program or domain-specific transactional database formal structure may impair the consistency of the warehouse database structure and the linkages with data from other sources.

## Data Analytics

Great effort has been placed into researching and developing analytical tools sitting on top of data warehouses to facilitate the generation of knowledge and intelligence. Open-source analytical digital applications (OSADA) and BI computer-based tools are an accessibility layer sitting on top of a data warehouse or similar data store. They make the underlying data more useful through better reporting, improved support for ad hoc data analysis, and predictive analytics.

- As data analytics market has grown, computerized analytic methods and tools have progressively become more sophisticated and applications that help users look farther along the decision-making processes are replacing simple retrospective analytics, which only displays a picture of the past.
- BI software application categories comprise data management tools, data discovery applications, and reporting tools including dashboards and visualization software. The trend in the BI market is to deliver a user-oriented software architecture and design making them easily employed by business users in analyzing particular datasets.

- The data analytics capabilities of BI products consist mostly of various aggregate operations (sums, averages, and so on) that are typically supported by OLAP (Online Analytical Processing) "cubes". OLAP "cubes" are multidimensional contingency tables; i.e., with more than two dimensions, that are precomputed and stored in the data warehouse allowing for relatively quick summaries or projections along any of the axes. These "cubes" behave much like spreadsheets and thus are familiar and accessible to individuals comfortable with spreadsheets and pivot tables.
- The analytic component allows downloading of results to be used by internal and external data visualization applications; data analysis and statistical tools (e.g., SPSS, Stata, SAS); MS Office applications (MS Excel, MS Access); and a variety of OSADA and report design applications. Visualization software allowing connection to the data warehouse provides an interface for exploring large multidimensional databases permitting intuitive drag-and-drop combination of multiple views of data.
- Some analytical resources include predictive analytics, which use past trends to help organizations understand future probabilities, and prescriptive analytics, which use past trends and data about the current state to recommend decision-making actions (Table 2.1).

Today, there are over hundred BI software companies selling some type of BI tool. At the Global Leadership Conference (GLT) meeting called by the World Health Organization in Geneva, Switzerland, on June 18−20, 2014, discussions were held regarding the strategic direction in the marketplace for BI products. The value and limitations of BI applications and analytics were examined including business and technological priorities and deployment issues.

**Table 2.1** Definition, Function, and Output of Different Classes of Business Intelligence Analytics

|  | Retrospective Analytics | Predictive Analytics | Prescriptive Analytics |
|---|---|---|---|
| Definition | Displays a portrait of what happened | Informs how future behavior might unfold based on current trends | Suggests actions to improve efficacy |
| Function | Requires interpretation by user | Trends already interpreted and predictions made requiring the user only to determine possible actions | Determines optimal responses to trends and guides user to high-value actions |
| Output | Typically are spreadsheets and dashboards and may have alerts when deviation from trends occur | Delivered alerts and action triggers within the application | Options presented to user in the form of menus with alerts recommending best action |

IMS, 2014. Riding the Information Technology Wave in Life Sciences: Priorities, Pitfalls and Promise. IMS Institute for Healthcare Informatics, Parsippany, NJ.

- The GLT looked at trends of BI products and examined in detail the characteristics of commercial applications. The following BI software market trends were noted: simplicity of use, flexible data discovery and analysis tools, mixture of technologies, off-the-shelf data models for specific applications, and a variety of graphic display tools. Investment recommendations emphasized the need for analytical skills development and appropriate data management and quality.
- The very high incidence of failure of BI corporate projects was highlighted—the most common reasons being (1) lack of engagement and adoption by intended users; (2) poor vision or strategy for the project; (3) data quality issues; (4) lack of agreement on data definition standards; (5) organizational culture and political issues; (6) inadequate governance; (7) creating awareness and ensuring adoption; (8) using the right metrics; (9) over-reliance on tools and technology; and (10) lack of information technology and analytical skills.

## Reporting

In an enterprise environment, reporting is the primary product of analyzed data. It is also one of the greatest sources of frustration. Given the ubiquity of reporting and the resources spent on it, one would think that the area would be pretty well understood and supported. However, reporting is a software application area barely touched by the notion of agility and user-friendliness.

- Reporting solutions are invariably big, bulky, bureaucratic, slow to change, and awkward to use. Attempting to conserve resources, reporting solutions are often built generically—a single reporting system that supports all the needs of all the users with most reporting confusing "up to date" with "real time."
- Clear distinction must be made between internal operational reports and representative reports, the latter intended for external users. Operational reports need to be fast and convenient. Representative reports need to be settling and visually appealing. It is not realistic to expect a single reporting system to support both requirements simultaneously.

## GENERATION AND USE OF KNOWLEDGE AND INTELLIGENCE

An organizational perspective for the best use of knowledge and intelligence requires first examining the intellectual and structural capabilities of an organization that require eventual reform, development, and fostering to encourage performance, adaptation, and innovation. Corrective interventions emphasize the steps organizations must take to stimulate closer connections among their decisions, operations, and emerging knowledge. The perspective posits two interrelated principles: capabilities and processes.

- When knowledge is endogenous, the challenge is to ensure its diffusion to other organizational units. When knowledge is exogenous, the challenge is to rapidly capture and translate it into innovative practices and services within the organization.
- The issue of underperformance of experts and organizations with respect to knowledge and intelligence is surprisingly frequent and a large body of research on learning organizations and evidence-informed management sheds light on how organizations generate and share those intellectual resources. They address issues of receptive capacity, learning, and involvement in the creation and coproduction of knowledge and intelligence.
- An organization can improve its ability to manage knowledge if its structure, strategy, and culture effectively support autonomy, flexibility, decentralization, group incentives, and knowledge brokering.

Knowledge and intelligence are dynamic and ambiguous intellectual constructs characterized by fluid boundaries requiring educational, social, and personal commitment that determine acceptability and understanding of their potentialities and limitations. It has been shown that knowledge and intelligence are adopted and effectively used when (1) they contribute to the improvement of individual problem-solving capacity; (2) increase the sense of self-control of the work environment context and day-to-day practice; and (3) they reflect normative preferences of what is expected an innovation should accomplish.

- The capacity to capture knowledge, generate intelligence, translate them into actions, and to learn from experience rests squarely on the behaviors, talents, and intellectual capacities of individuals.
- The concept of knowledge as a set of capabilities holds that organizations will excel in knowledge management if they manage the tension between the autonomy of a decentralized structure and the need to stimulate professionals to improve their performance.
- Social processes that support the constitution and circulation of knowledge in organizational networks are crucial determinants of use and application. The concept of evidence-based decision is contingent on the ability of people within an organization to agree on a common set of problems and to maintain cooperation and communication despite inevitable controversies that will certainly arise.
- Within health organizations, it is common practice to delegate technical and operational autonomy to experts with the aim of

applying up-to-date knowledge in the solution of health issues. However, in relation to information systems, autonomous and highly qualified professionals do not guarantee the quality and safety of data and information generation, nor the analytical work needed to produced intelligence and its application to the updating of practices to cutting-edge knowledge and technologies.

- Ideally, decision supported by strong intelligence—i.e. evidence-based—is the preferred way to implement knowledge although decisions can be, and frequently are, made at any level of the value-added chain progressing from data to information and then to intelligence. Experience, however, has shown that the appropriateness of a decision is enhanced as one move along the chain.

How the value-added continuum *data > information > knowledge > intelligence* operates in practice is illustrated by the processes related to how therapeutic interventions such as drugs or medical procedures are discovered, tested, and introduced in clinical practice—from their early research and identification as promising candidates for adoption, data are gathered and accumulated as information regarding their nature, utilization, and effectiveness. The expansion of the knowledge about the associated particulars related to a specific therapeutic agent, its desirable and unwanted effects and interactions are further acquired over time thus building a body of knowledge (knowledge base). Finally, collaborative peer-reviews of existing knowledge are conducted and published in the scientific literature or as collaborative reviews (e.g., Cochrane Collaboration Reviews) establishing an evidence-base foundation for protocols of use (intelligence).

## COMMUNICATION AND WEB PUBLISHING

NHIS and HOs must communicate data, information, knowledge, and intelligence to potential users by targeting, tailoring, and packaging contents to dissimilar audiences. Strategies to accomplish those objectives include:

- Developing a user-oriented dissemination strategy taking into account the promotion of information/knowledge/intelligence user networks and social communication media engagement.
- Establishing a web-based technological infrastructure to share relevant analyses in a variety of interactive tabular, graphic, and textual formats.
- Dissemination processes and approaches should include context-specific evidence including, if available, references to systematic reviews developed by the issuing organization or through links to the literature.

- Shared content should be clear and action-oriented with careful consideration given to expected effect upon audience decision-makers.
- Availability of high—quality, up-to-date sources containing statistics and metadata with different levels of granularity and disaggregation—from country to local level as required by the type of information under examination.
- Transparency by publishing the methodology used and a description of data adjustment and estimation together with the peer-reviewed routines employed in the analytical work.
- Support by an advisory group made up of institutional and external experts in the domain about which the information is being provided to assist in the development of data validation rules and estimates, analytical methodology, and independent review.
- Attract individuals with diverse experience, marketing and social communication skills, and the capacity to span boundaries and understand the potentially disparate worlds of researchers and knowledge users. They are ideal knowledge brokers and may advantageously serve as (1) intermediaries to facilitate collaborations between researchers and knowledge users; (2) helpers in the identification of the evidence required for decisions; (3) assess, interpret, and adapt evidence to local context and recognize emerging issues.
- Collaborative relationships between researchers and knowledge users built on trust and frequent interaction enhance the effectiveness of dissemination and exchange activities. Professional social networks and communities of practice hold promise as innovative ways to disseminate and exchange knowledge.
- A web public accessed portal with visual display and navigation aids as the single point of entry for (1) dissemination of data, information, knowledge, and intelligence; (2) selective access to data, statistics, information, and analytical results; (3) technical and methodological recommendations; (4) summaries of the health situation and trends, and (5) products offered by the NHIS or HO to authorized users.
- Web Portal with a minimum of three main components for dissemination of information: (1) database/repository that includes easy links to all major databases; (2) analytical reports on specific and crosscutting topics; and (3) pages providing access to statistics, reports for geographically defined locations (country, region, subregion, administrative or political unit, local).
- Monitoring of the public portal audience distinctiveness and expectations by portal utilization metrics and being ready to make appropriate changes of the web interface design and features.

A fully fledged Web Portal depends on an advanced data processing and analytical infrastructure with capacity of supporting the generation and visualization of:

- Analytical reports on the current situation and trends for priority health issues. These include the production of crosscutting statistical reports as well as statistical reports from each specific program.
- Access to an interactive data repository containing health statistics comprising indicators, metadata, and dynamic table views with filter, pivot, and download functionalities allowing users to view and filter core indicators of different health topics for countries and regions, and download the customized tables in plain text, PDF, or Excel formats.
- Dashboards providing a high-level view of all relevant metrics in a single report rather than a collection of individual detailed reports.
- Links to theme pages, geographic and demographic information, and other internal and external information products.
- Theme pages highlighting health priorities organized by grouping core indicators and providing analytical views of the data in predefined graphic outputs including dashboards, charts, and static and interactive maps.

## EXPERIENCE WITH NHIS AND HOs

### National Health Information Systems

Existing NHIS display a great variety in terms of implementation design, operational routines, resources, performance, and effectiveness. That diversity depends directly on the level of development reached by a country and its healthcare system. Traditionally, NHIS were based on historical data collected and manipulated manually at central statistical offices, in order to generate highly aggregated statistics and periodic reporting. The spread of digital and telecommunications technologies and the work of international health organizations have led to the progressive automation and integration of standalone data processing systems at all levels of the health system.

Plans and projects for NHIS implementation and improvement have been applied to the nationwide system, as an integrated network under the coordination of Ministries of Health resulting in increasing capacity and accuracy in the production of information, even in less-developed countries. Concomitantly there has been an effort at strengthening specific mission-critical information systems, such as (1) hospital management information systems; (2) laboratory, pharmacy, and other support services; (3) epidemiological surveillance; (4) community health services; (5) health information network; (6) virtual health libraries, and others.

Results from investment and implementation efforts have been a mix of successes, limitations, and outright failures due to multiple causes not necessarily related to issues related to technology deployment. Systematic evidence indicates that many NHIS are poorly equipped to meet current information demands for policy decisions and are ill prepared to meet future requirements. NHIS typically have shown low integration between specific technical systems and administrative levels, resulting in limited information quality and overall systems performance. They are often (1) isolated from policy- and decision-making and nonresponsive to information needs; (2) with limited capacity to put in place and sustain technologically driven systems; (3) automation and technology is applied to information systems under restrictive conditions with institutional, statistical, and functional limitations; and (4) limited institutional culture for demanding, integrating, analyzing, and using information.

Assessment studies of NHIS conducted by a variety of researchers report a common catalog of constraints, especially in low-income and lower middle-income countries. Evaluation of NHIS in low and lower middle-income countries published in 2011 and 2012 by the World Health Organization HMN reported a long list of limitations that should be taken into account in the search for better NHIS implementations. Common findings reported by the WHO/HMN studies are:

- Lack of well-formulated and implemented health information systems policies. All assessed countries struggle with the need for well-formulated and implemented projects. The formulation of health information policies should directly involve the statistics and epidemiology units.
- Low priority is reflected in insufficient funding by health authorities and financial governmental sources. Limited availability of resources to implement and maintain the NHIS is a frequent cause of failure after initial deployment.
- Health administration units, hospitals, and health centers without appropriate human resources in terms of number and qualification to implement and operate systems at an acceptable level.
- Even though decision-makers invariably state that they need a well-functioning information system, health authorities are hesitant to allocate financial resources at district and regional level for the core activities related to collecting and processing of data.
- Basic concepts of data processing often are not understood. Health managers, in some instances, were not aware of the data and information required to manage and monitor their tasks efficiently.
- There are widespread expectations by health administrators that computers can solve their management problems with unrealistic hope

that computers will solve problems that are related to lack of infrastructure and governance.

- Health service personnel are frustrated by increasing paperwork. Most NHIS in the assessed countries used manually collected data. Duplication of the same data on different forms is very common. The paperwork filled out on a regular basis is generally burdensome resulting in wasted time for health personnel that rightfully perceive actual patient care as their core activity. Worse still, data are collected which are not processed.

- The need for and usefulness of data collection is not clearly understood by those involved. Validity of data suffers from uncertain procedures regarding how to fill out correctly a vast number of forms. Often, policy and procedure manuals are not available at the user level and, when they do exist, seldom enough emphasis is placed on adequate training of personnel on how to fill out the forms correctly and the purpose of the task.

- Lack of timeliness of data is a common issue. Data are summarized and forwarded to the referral level months after collection, therefore not in time for practical use by health planners and administrators.

- Training of personnel from statistics departments is essential to run highly complex health information systems. Frequently there is high turnover and attrition of personnel and a constant training scheme is necessary.

- There is a serious lack of accurate catchment population data. Frequently exact population data are not available due to missing or incomplete civil registers or adjusted population figures are not calculated by using the latest available data. Particularly, for the calculation of coverage of antenatal care and vaccination, it is mandatory to have data about the catchment population of each community health service to be able to identify vulnerable populations.

- The information products provided by the system are questionable in their consistency, data collection method, representativeness, and periodicity. End-users must have clear job instructions, clear management roles, and well-defined program goals with set targets and indicators to measure target achievement.

- Data and information not used for decision-making. The information system must clearly identify what are the proper requirements at the most critical decision-making levels. Health managers are the key persons to decide what data has to be collected and for what purpose. A complete involvement of these key persons during the development of the system is a prerequisite to ensure its utilization.

- Lack of interoperability. Existing information systems often are not compatible making collecting and access to data cumbersome or even impossible. A national health data warehouse was not available in any of the low-income and lower middle-income countries reported in the WHO/HMN studies.

- Software and hardware need periodic maintenance, and that imply its own infrastructure or even a new unit with its proper financing.
- Retaining qualified information technology personnel is a challenge with the low salary schemes at the Ministries of Health vis-à-vis the opportunities offered by the private commercial, industrial, and financial sectors where the demand for those professionals is high.

## Health Observatories

PHOs have been implemented at international, national, and subnational level. Examples of successful experiences are found in the Association of Public Health Observatories (APHO), involving a network of PHOs working across European countries. APHO produces data, information, knowledge, and intelligence on health and healthcare used by practitioners, policy-makers, and the wider community.

The World Health Organization implemented the Global Health Observatory (GHO) and the Pan American Health Organization (PAHO/WHO) established the Regional Health Observatory for the Americas, aligned with the model and contents of the GHO. The WHO Regional Office for Europe (EURO/WHO)—in partnership with some governments and agencies—has developed the European Observatory on Health Systems and Policy, which has been producing relevant systematic information and technical reports on health systems and health policy at national and regional levels.

A HO makes greater use of the statistical dimension of health situation displaying trends, geographic variations, digital maps, and graphs. A statistical model with indicators, tables, maps, graphs, and some brief specific comments are found in most HO such as the WHO GHO and the PAHO Regional Health Observatory. At national and subnational levels, HOs tend to focus on the monitoring and surveillance of specific health-related areas.

Common issues covered by observatories include (1) health situation of communicable and noncommunicable diseases; (2) healthcare systems' performance; (3) human resources; (4) impact of health sector reforms; (5) public health surveillance; (6) natural disasters and preparedness; (7) surveillance of inequities and determinants in health; (8) occupational health; (9) accidents and violence; and (10) ethnical and gender issues. Some implementations have integrated a large number of datasets and are useful policy-oriented resources, as is the case of the European Observatory for Health Systems and Policies and its related Health Policy Monitor.

A HO is functionally consolidated when it can perform all its expected functions, including surveillance of the health situation and determinants—the "demand side"—and monitoring of the health sector—the "supply side"—followed by integrated contextual analysis and forecasting, with production of reports that are policy-oriented and focused on the needs and culture of decision-makers.

The development of HOs is facilitated by factors such as:

- Improvement in policy and decision-making leading to increased recognition, demand, and use of information.
- Perceived needs, opportunities, and experience of countries and agencies in establishing better accountability, monitoring, and evaluation routines to strengthen health systems performance and attainment of planned outcomes.
- International collaboration and networking between different participating entities around public health with the support of international organizations as is the case of the World Health Organization.
- International agreements on criteria and standards to improve harmonization of information allowing comparison and consolidation.
- Support provided by advanced resources of information and communication technologies, facilitating the effective production, integration, and updating of large quantity of information at national and international levels.

The experience of implementing HOs, suggests that success is chiefly related to (1) a design that is responsive to national, regional, and local information priorities and (2) taking into account feasibility, capacity, and readiness to shift from current fragmented information, surveillance, and monitoring systems toward an integrated and effective model. Moreover, successful HOs were found to receive adequate support from similarly well-run NHIS since:

- Most data and information needed for the operation of an observatory is already available in different national information sources even when that information is not formally reported or published such as contents provided by the nationwide health surveillance and health system monitoring applications.
- Some HOs integrate all the relevant information produced by the NHIS and other sources that provide health-related information. In some instances, HOs even replicate portions of the NHIS role however, the analytical component, visualization tools, and reporting provide a more comprehensive and opportune panorama.
- Following international agreements and recommendations, some countries have developed useful monitoring systems, such as those

focused on the Millennium Development Goals (UN), and the Observatory of Human Resources (PAHO/WHO) and Health Sector Reform (PAHO/WHO).

- Staff working on the development and updating of national health profiles and health systems profiles have the capacity to perform descriptive and analytical overview.

## IMPLEMENTATION OPPORTUNITIES AND CHALLENGES

### National Health Information Systems

Improvement of NHIS, in support to policymaking and monitoring of evaluation of national strategies and plans, is of great interest and focal point of action of international and intergovernmental development agencies and partnerships. Decisions for investing in the NHIS—to improve information and its use—are justified by the soaring need for information for health systems management and operation as long as encouraging conditions of feasibility, readiness, and cost benefit do exist. Implementing and strengthening NHIS involves enhancing the technical, administrative, technological functional components, and operational resources.

The Global Health Information Forum 2010, with the participation of governmental, nongovernmental, multilateral, and international agencies; and development partner organizations, made a call to action with consensual agreement to the following general principles:

- Transparency of all health data that considers privacy and confidentiality issues while providing unhindered access to those who are the sources or originators and to end-users of data and information.
- Good governance, so as to ensure that (1) information is collected from public and private healthcare providers; (2) national health strategies are based on evidence provided by information systems; (3) national performance monitoring tools are implemented to track health goals; and (4) monitoring health equity, and strengthening intersectoral collaboration and coordination on NHIS is implemented.
- Investments and capacity building allocating at least 5% of health budgets to NHIS and appropriate technologies.
- Harmonization and integration of data produced by national statistical offices, ministries of health and other sectors and facilitation of intersectoral collaboration, data sharing, and community participation and ownership.
- Planning for future expansion of appropriate technological solutions within and among countries.

The WHO/HMN was established on the premise that better health information means better decision-making, leading to better health. The HMN goal is to increase the availability, quality, value, and use of timely and accurate health information. HMN proposes a set of principles to guide the NHIS development, including aspects such as:

- Build upon existing initiatives, systems and knowledge with broad-based consensus and stakeholder involvement. Favorable prerequisites will depend on (1) extant national and multisector capacity to produce information; (2) the adoption of policies and common criteria, standards and compatible databases; (3) multisector and multi-agency support; (4) continuous political support; (5) financial, human, and technological resources; and (6) establishment of a technical national group with a steering and managerial role to guide, implement, and manage the system.
- Management of improved NHIS is facilitated if the system has been well designed in accordance with country information needs and responsive to user demands. Opportunities for implementation and consolidation should focus on key aspects, such as specific information systems, sources of information, participant technical and management units, administrative levels, and geographic regions.
- Planning and improvement of NHIS should consider the national capacity for the management and sustainability of the NHIS. It is essential to adopt and keep clear rules, responsibilities, operational criteria, and standards for the health information system. NHIS administrative improvement covers the legal and administrative framework including policies, norms, regulations, organization, and the management of resources and processes.
- Improve national capacity, leadership, and ownership. To be successful, efforts made to improve structure and performance must consider the national capacity and the feasibility to enhancing critical aspects (staff training, methods, resources, technology, information flow) that can limit the quantity, quality, and coverage of information.
- Developing and strengthening the NHIS should be a gradual, incremental process with clear design and planning, administrative arrangements, and training including technical norms and guidelines on participants' responsibilities and the procedures to be followed
- Appropriate support of connectivity resources and services provided by the information and communication technological platform including data security, service continuity, user support, and regular application upgrading.

External technical assistance and the mobilization of resources are important, especially if regular budgets assigned to health information systems are limited and cannot support investments toward better information systems. In January 2010 the Health 8 (H8), an informal group of 8 health-related organizations comprising WHO, UNICEF, UNFPA, UNAIDS, GFATM, GAVI, Bill & Melinda Gates Foundation, and the World Bank called for the need to strengthen key data sources and capacity for analysis, synthesis, validation, and use of health data in countries. With that goal in mind, those agencies proposed to (1) increase the level and efficiency of investments in health information; (2) develop common data architecture; (3) strengthen performance monitoring and evaluation; and (4) increase data access and use.

## Health Observatories

The HO must be a policy-oriented and managerial tool responsive to the information needs of decision-makers. They should be relevant actors at all stages of planning, implementation, and management. International organizations—such as PAHO/WHO, EURO/WHO, and the WHO HMN—have been promoting harmonized and standardized datasets, guidelines, methods, software, and experience to develop, collect, and analyze health-related data. The overall contribution expected from those initiatives can cover most of what is necessary for an effective collection, management, and analysis of relevant information.

After the development of a HO design and plan, careful project monitoring and periodic evaluation to ensure its progressive development and consolidation should follow the implementation process. Successfully implemented national and international observatories, such as those participating in the APHO and the European WHO Observatory for Health Systems and policies have become good empirical models for the design and implementation of HOs elsewhere. Main aspects to be taken into account are:

- Level and continuity of political and institutional interest and effective support to the proposed observatory.
- Political and financial feasibility of planning and implementing the HO on a short-, middle-, and long-term timeframe.
- Assessment of infrastructure, user expectations, as well as the opinion of experts regarding the feasibility of developing the components of the observatory at short, medium, and long term.
- National capacity and competences in disciplines such as public health, epidemiology, health policy, planning, and decision-making.
- Managerial culture for information search, use, analysis, and application on decision-making.
- Diagnosis of the information needs of the HO's main potential users, especially national authorities, with emphasis on those responsible for

policy- and decision-making, as well as their health data analysis and communication advisors.

- Existence of historically stored information and databases containing core health-related data and other key health-related information.
- Extent of definition and standardization of the necessary key data.
- Quality, timeliness, completeness, and validity of each type of relevant data and indicator.
- Existence and effectiveness of performance measurement for each specific surveillance and monitoring system.
- Existence or potential establishment of a nationwide and multientity health information network.
- Level and quality of information sharing among different sources of information and role of the national health authority to collect and integrate it.
- Existence or affordability in the deployment of the most appropriate information and communication technologies platforms aimed at a health information network that is responsive to the national information needs.

## Further Reading

Anderson, M., Cosby, J., Swan, B., Moore, H., Broekhoven, M., 1999. The use of research in local health service agencies. Soc. Sci. Med. 49, 1007–1019.

CIHR, 2004. Knowledge Translation Strategy 2004–2009: Innovation in Action. Canadian Institutes of Health Research, Ottawa, ON, Canada, Cat. No.: MR21-56/2004E-HTML. ISBN: 0-662-38785-6.

Dalkir, K., 2005. Knowledge Management in Theory and in Practice. Butterworth-Heinemann Publisher, London, UK.

Eng, T.R., Gustafson, D.H., 1999. Wired for Health and Well-Being: The Emergence of Interactive Health Communication.. Science Panel on Interactive Communication and Health, Office of Disease Prevention and Health Promotion, US Department of Health and Human Services, U.S. Printing Office, Washington, D.C, April.

Few, S., 2006. Information Dashboard Design: The Effective Visual Communication of Data. O'Reilly Media, Inc, Sebastopol, CA, 978-0-596-10016-2.

Floridi, L., 2010. Information: A Very Short Introduction. Oxford University Press, Oxford, 152p, 978-0-19-955137-8.

Gattini, C., 2007. Investing in better health information systems: strategic recommendations for effective implementation and maintenance. Health in Regional Development. CARICOM, Guyana.

Gattini, C., 2009a. Improving the Structure and Performance of National Health Information Systems: Operational Approach and Strategic Recommendations. Pan American Health Organization/World Health Organization Country Office, Chile, Santiago, p. 118.

Gattini, C., 2009b. Implementing National Health Observatories: Operational Approach and Strategic Recommendations. Pan American Health Organization/ World Health Organization Country Office, Chile, Santiago, p. 65.

Health Metrics Network, 2005a. Health Metrics Network: What it is, what it will do, and how countries can benefit? Document Draft 1.0 August 2005. World Health Organization, Geneva.

Health Metrics Network, 2005b. Strengthening Country Health Information Systems: Situation Analysis Tool. Document Draft 1.0 August 2005. World Health Organization, Geneva.

Health Metrics Network, 2005c. Towards a Framework and Standards for Country Health Information System Development. Version 1.0. World Health Organization, Geneva.

Health Metrics Network, 2005d. Issues in Health Information 1: National and Sub-national Health Information Systems. (Unedited). World Health Organization, Geneva.

Health Metrics Network, 2006. Issues in Health Information 4: Improving Systems for Measuring and Monitoring Vital Events. (Unedited). World Health Organization, Geneva.

Health Metrics Network, 2007. Framework and Standards for Country Health Information Systems. second ed. World Health Organization, Geneva.

Health Metrics Network, 2010. The Global Health Information Forum 2010: Summary Report. Bangkok, Thailand 27–30 January 2010. World Health Organization, Geneva.

Health Metrics Network, 2011. Country Health Information Systems: A review of the current situation and trends. World Health Organization, Geneva, ISBN 978 92 4 156423 6.

Health Metrics Network, 2012. Country Health Information Systems Assessments: Overview and Lessons Learned. Working Paper 3. November 2012. World Health Organization, Geneva.

Hemmings, J., Wilkinson, J., 2003. What is a public health observatory? J. Epidemiol. Commun. Health 57, 324–326.

IMS, 2014. Riding the information technology wave in life sciences: priorities, pitfalls and promise. March 2014 IMS Institute for Healthcare Informatics. Parsippany, NJ.

Jadad, A.R., Cook, D.J., Jones, A., Klassen, T.P., Tugwell, P., Moher, M., et al., 1998. Methodology and reports of systematic reviews and meta-analyses: a comparison of Cochrane reviews with articles published in paper-based journals. JAMA 280 (3), 278–280.

Janert, P.K., 2011. Data Analysis with Open Source Tools. O'Reilly Media, Inc, Sebastopol, CA, 978-0-596-80235-6.

Kaegi, L., 1998. Health services researchers zero in quality, policy, and practice in a changing health care system: report of the 15th Annual Meeting of the Association for Health Services Research. Jt. Comm. J. Qual. Improv. 24 (11), 658–671.

Kleinau, E., 2000. Management of health information systems. Design and Implementation of Health Information Systems. WHO, Geneva, pp. 176–197. (Chapter 10).

Kothari, A., Rudman, D., Dobbins, M., Rouse, M., Sibbald, S., Edwards, N., 2012. The use of tacit and explicit knowledge in public health: a qualitative study. Implement. Sci. 7 (1).

Lippeveld, T. (n.d.). Routine Health Information Systems: The Glue of a Unified Health System. Vice-President of International Division, John Snow Inc. The RHINO Workshop on Issues and Innovation in Routine Health Information in Developing Countries, March 14–16, 2001, The Bolger Center Potomac, MD, USA.

Lippeveld, T., 2000. The context of health information system reform. Design and Implementation of Health Information Systems. WHO, Geneva, pp. 225–242. (Chapter 13).

Lippeveld, T., Sapirie, S., 2000. Approaches to strengthening health information systems. Design and Implementation of Health Information Systems. WHO, Geneva, pp. 243–252. (Chapter 14).

Lippeveld, T., Sauerborn, R., 2000. A framework for designing health information systems. Design and Implementation of Health Information Systems. WHO, Geneva, pp. 15–32. (Chapter 2).

Lyons, R., Warner, G., Langille, L., Phillips, S.J., 2006. Piloting knowledge brokers to promote integrated stroke care in Atlantic Canada. Evidence in Action, Acting on Evidence: A Casebook of Health Services and Policy Research Knowledge Translation Stories. Canadian Institutes of Health Research, Ottawa, ON, Canada.

Minkler, M., 2005. Community-based research partnerships: challenges and opportunities. J. Urban Health 82 (Suppl. 2), ii3–ii12.

Mitton, C., Adair, C.E., McKenzie, E., Patten, S.B., Perry, B.W., 2007. Knowledge transfer and exchange: review and synthesis of the literature. Milbank Q. 85 (4), 729–768.

NCDDR, 2005. What is Knowledge Translation? Focus, Technical Brief No. 10. National Center for the Dissemination of Disability Research. National Institute on Disability and Rehabilitation Research (NIDRR), the U.S. Department of Education's Office of Special Education and Rehabilitative Services (OSERS), Austin, TX.

Organization of American States, 2003. Telehealth in the Americas. Inter-American Telecommunication Commission (CITEL), in collaboration with the Telecommunication Development Bureau of the International Telecommunication Union (ITU), the Pan American Health Organization (PAHO/WHO), and the Latin American Association of Research Centers and Telecommunication Enterprises (AHCIET). Publication OEA/Ser.L/XVII.6.4, December 16, 2003, Washington D.C., 56 p, ISBN 0-8270-4614-6.

Robinson, T.N., Patrick, K., Eng, T., Gustafson, D., the Science Panel on Interactive Communication and Health, 1998. An evidence-based approach to interactive health communication: a challenge to medicine in the information age. JAMA 280 (14), 1264–1269.

Rodrigues, R.J., 2000. Information systems: the key to evidence-based health practice. Bull. World Health Org. 78 (11), 1344–1351.

Rodrigues, R.J., 2003. Opportunities and challenges in the deployment of global e-Health. Int. J. Healthc. Technol. Manage. 5 (3/4/5), 335–358.

Rodrigues, R.J., 2008. Compelling Issues for Adoption of e-Health. The Commonwealth Health Ministers Reference Book 2008. Henley Media Group Limited. Commonwealth Secretariat, Marlborough House, Pall Mall, London, pp. 36–40. ISBN 97 80955 44086 1.

Rodrigues, R.J., Gattini, C., Almeida, G., 1999. Setting Up Healthcare Services Information Systems: A Guide for Requirement Analysis, Application Specification, and Procurement. Pan American Health Organization Essential Drugs and Technology Program, Division of Health Systems and Services Development. PAHO/WHO, Washington, DC, ISBN 92 75 12266 0.

Straus, S., Tetroe, J., Graham, I.D., 2011. Knowledge Translation in Health Care: Moving from Evidence to Practice. BMJ Books, Wiley-Blackwell, UK, 978-1-4443-5725-7. 336 p.

UNCTAD, 2008. Information Economy Report 2007–2008: Science and Technology for Development—The New Paradigm of ICT. Overview Version. United Nations, New York and Geneva.

University of North Carolina (Chapel Hill), 2011. Measure evaluation: tools for data demand and use in the health sector. Performance of Routine Information Systems Management (PRISM) Tools. University of North Carolina at Chapel Hill, Chapel Hill, NC.

Wagner, C., 2006. Breaking the knowledge acquisition bottleneck through conversational knowledge management. Inform. Resour. Manage. J. 19 (1), 70–83.

WHO, 2010. Global Health Observatory—Steps and Procedures for WHO Programmes (Draft Protocol), May 2010. WHO, Geneva.

WHO, 2011. Monitoring, Evaluation and Review of National Health Strategies: A Country-Led Platform for Information and Accountability. 10 November 2011. World Health Organization and International Health Partnership, Geneva, ISBN 978 92 4 150227 6.

# CHAPTER 3

# Information and Communication Technologies and Global Health Challenges

**M.A. Gutierrez, R.A. Moreno and M.S. Rebelo**
Heart Institute, Clinics Hospital, University of Sao Paulo Medical School, São Paulo, Brazil

## CONTENTS

## INTRODUCTION

Information and communication technology (ICT) is an extended term for information technology (IT) which stresses the role of unified communications and the integration of telecommunications infrastructure (telephone lines, cable networks, wireless signals), computers and software. ICT enables users to access, store, transmit, and manipulate data. The term ICT is also used to refer to the convergence of audiovisual and telephone networks with computer networks through a single cabling or link system. However, ICT has no universal definition, as the concepts, methods, and applications involved in ICT are constantly evolving on an almost daily basis. The broadness of ICT covers any product that stores, retrieves, manipulates, transmits, or receives information electronically in a digital form.

Over the past 15 years the ICT revolution has driven global development in an unprecedented way. Technological progress, infrastructure deployment, and decreasing prices have brought an unexpected growth in ICT access and connectivity to billions of people around the world. According to the International Telecommunication Union, in 2015 there were more than 7 billion mobile cellular subscriptions worldwide, up from less than 1 billion in 2000. Globally, 3.2 billion people were using the Internet, of which 2 billion were from developing countries (ITU, 2015). This rapid development of ICT offers great opportunities for many sectors to improve efficiency and reduce costs, but at the same time, it poses new challenges. While the population can access more information in real time, important points need to be addressed: legal issues, such as privacy and security; and social problems, such as the digital divide.

**Global Health Informatics.**

Healthcare is mainly an information business, and the quality, efficiency, and outcomes of care depend on effectively capturing and managing patient information. There is no healthcare without management, and there is no management without information. Despite of different approaches to the organization and financing of healthcare, countries across the globe are moving toward increased use of ICT to improve healthcare delivery. There is a shared enthusiasm about the potential for these technologies to address the common challenges of inconsistent quality and inefficient delivery of care. There is also a growing sentiment that health ICTs are essential to reform existing care delivery models and to change rapidly the way care will be provided in the future (Jha et al., 2008; Mechanic, 2008; Bates and Bitton, 2010).

In this chapter, we will cover five areas of ICT in healthcare: electronic health records (EHRs), Health Information Exchange (HIE), telemedicine, mHealth, and social media. In each of these areas, we examine the current applications and discuss the main challenges to reach a wider adoption. Steps to better EHRs include review the current health record system, try to emulate benchmark practices, involve the anticipated users of the system from the onset of discussions, train the users to the EHR system, evaluate the benefits of the implemented system, and update the system when needed. HIE requires the simplification and harmonization of existing standards, more precise legal rules for healthcare exchange, subsidies to sustainability. Telemedicine needs legal definition for providing and consuming services, subsidies to sustainability, capable of adapting to different needs of each region. mHealth has to overcome security issues, integrate to national and international healthcare programs and standards, define metrics for quantifying for its real benefits, subsidies to sustainability. Social media needs to deal with the problem of reliability of information, privacy and confidentiality, definition of guidelines for healthcare professionals, more precise legal rules for regulating the health communications.

## ELECTRONIC HEALTH RECORDS

Hospitals began investing in health ICTs in the 1960s, and since then they have made big investments for the development of various types of ICT applications. ICTs were first used to support auxiliary functions, such as financial services; in a later phase, ICTs were utilized to manage pharmacy, laboratory and radiology service lines, thus to monitor and support clinical activities (Agha, 2014).

In 1991 The Institute of Medicine (IOM) defined the term computer-based patient record (CPR) as an electronic patient record that resides in a system specifically designed to support users by providing accessibility to complete and accurate data, alerts, reminders, clinical decision support systems, and links to medical knowledge (Institute of Medicine, 1991). Since the publication of this report, a variety of terms has been used in the literature for ICTs applications to support clinical activities. By the late 1990s the term CPR was replaced by the terms Electronic Medical Record (EMR) or EHR. In 2008 the National Alliance for Health Information Technology proposed standard definitions for the EMR, the EHR, and the personal health record (PHR). Table 3.1 presents the summaries of these definitions (Duke et al., 2006).

The IOM (Institute of Medicine 2003) defines the EHR as a system that can perform eight electronic functions. According to this definition the four core functions of an EHR are related to a single healthcare organization. The other functions refer to systems that share information across different organizations or health information organization (Table 3.2).

An EHR is a digital version of a patient's medical history. As such, it must be able to electronically collect and store patient data, supply that information to providers on request, permit clinicians to enter orders directly into a computerized provider order-entry system, and advise healthcare professionals by providing decision support tools such as reminders, alerts, and access to the latest research findings, or appropriate guidelines. It is a longitudinal record of patient health information generated by one or several encounters in any healthcare providing setting. It includes a full range of data relevant to a patient's care such as demographics, problems, notes, medications, observations, medical history, immunizations, laboratory data, radiological images, and vital signals and its reports.

The development of EHR has greatly expanded the automation of clinical services. These systems have the ability to integrate information from pharmacy, radiology, and laboratory in a way that allows physicians and other health professionals to directly access this information and have a complete and integrated picture of a patient. Fig. 3.1 shows a typical interface of an EHR system that includes a complete view of patient clinical information, including text, signals, and images (Furuie et al., 2006).

**Table 3.1** Health Information and Communication Technology definitions (Duke et al., 2006)

| Electronic Medical Record (EMR) | Electronic Health Record (EHR) | Personal Health Record (PHR) |
|---|---|---|
| An electronic record of health-related information on an individual that can be created, generated, managed, and consulted by authorized clinicians and healthcare professionals in one healthcare organization. | An electronic record of health-related information on an individual that conforms to nationally recognized interoperability standards and that created, generated, managed, and consulted by authorized clinicians and healthcare professionals across more than one healthcare organization. | An electronic record of health-related information on an individual that conforms to nationally recognized interoperability standards and that can be drawn from multiple sources while being managed, shared, and controlled by the individual. |

**Table 3.2** Functions of an EHR System According to the Institute of Medicine

| Core Functions | Other Functions |
|---|---|
| **1.** *Health information and data* includes medical and nursing diagnoses, a medication list, allergies, demographics, clinical narratives, and laboratory test results. | **5.** *Electronic communication and connectivity* enables those involved in patient care to communicate effectively with each other and with the patient. |
| **2.** *Results management* manages electronically all types of results, such as laboratory test results and procedure results based on images and biomedical signals. | **6.** *Patient support* includes everything for patient education materials and home monitoring by means of telehealth. |
| **3.** *Order entry and support* incorporates the use of computerized order entry by the physicians for medication and exams (CPOE). | **7.** *Administrative processes* facilitates and simplifies processes as scheduling, prior authorization, insurance verification, decision support tools to identify eligible patients for clinical trials or chronic disease management programs. |
| **4.** *Decision support* employs computerized clinical decision support capabilities such as reminders, alerts, and computer assisted diagnosing. | **8.** *Reporting and population health management*establishes standardized terminology and data formats for public and private sector reporting requirements. |

*Source: (Adapted from Institute of Medicine 2003)*

The EHR systems have several advantages over the traditional paper-based methods to collect and register patient information. However the transition from paper-based health records to EHR in a healthcare setting takes time (Delpierre et al., 2004; Uslu and Stausberg, 2008). There are certain factors that influence this migration, which include the availability of financial support, uncertain return on investment, and standard of technology, level of resistance to change and level of priority for change.

## EHR for Epidemiologic Research

The use and functionality of EHRs have increased rapidly in the past decade. Although the primary purpose of EHRs is clinical, researchers have used them to conduct epidemiologic investigations, ranging from cross-sectional studies within a given hospital to longitudinal studies on geographically distributed patients (Sa et al., 2016). The use of EHRs to produce secondary databases for epidemiologic studies involves algorithms to data extraction, transformation, and loading from a variety of settings. The most successful EHR researches to date have used anonymized databases in the UK and US healthcare systems, whose patient populations receive most or all of their care within the system. In these cases the EHRs are used for comparative effectiveness and health services research, pharmacoepidemiology and genetics epidemiology, and disease surveillance.

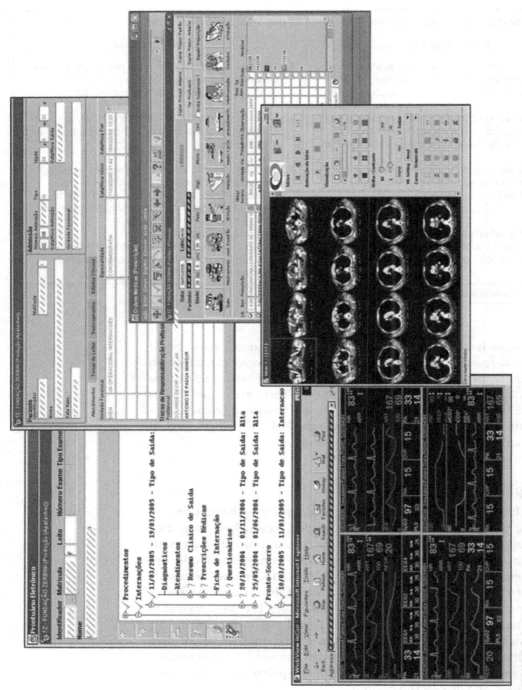

**FIGURE 3.1**

Typical interface of an EHR system that includes a complete view of patient clinical information, including text, signals, and images (Furuie et al., 2006).

*EHR*, electronic health record.

Kaiser Permanente in the United Staes has several EHR-based cohorts (Armstrong-Wells et al., 2009; Croen et al., 2005), including the Diabetes Study of Northern California (DISTANCE) study (Moorman et al., 2013). DISTANCE involves 20,000 patients with diabetes and has addressed wide-ranging issues, including diabetes outcomes among Asians and Pacific Islanders (Kanaya et al., 2011), the impact of neighborhood deprivation on cardio metabolic health indicators (Laraia et al., 2012), and the relationship of low social income status to risk of hypoglycemia (Berkowitz et al., 2014).

The use of EHR information from different health systems can also be observed in the assembly of multisystem cohorts. In this area the HMO Research Network has been a leader in this type of research since 1994 (Selby, 1997). Other examples are: the Consortium on Safe Labor (Hibbard et al., 2010; Männistö et al., 2015), which uses EHR delivery and birth data from 19 hospitals; the Clinical Assessment, Reporting, and Tracking system in Veterans Administration hospitals (Vigen et al., 2013); and the Chronic Hepatitis Cohort Study, which combined data from four healthcare systems on more than 1.6 million adults to identify a cohort of patients with hepatitis B and hepatitis C (Moorman et al., 2013). Studying Chronic Hepatitis Cohort data, Mahajan et al. (2014) found that only 30% of hepatitis C−positive patients who died with documented liver disease had hepatitis C on their death certificate, uncovering huge underestimates of the role of hepatitis C on mortality in the United States.

Around the world, there are initiatives to study populations from central repositories of anonymized data generated from EHRs. Examples come from the Clinical Data Analysis Report System in Hong Kong (Cheuk et al., 2004) and the Clinical Practice Research Datalink (Hesdorffer et al., 2012; Qizilbash et al., 2016), the combination of the General Practice Research Database, one of the world's largest and best-established research databases of electronic primary care data, the QResearch, which holds data for more than 12 million patients, the Health Improvement Network (THIN) (Osborn et al., 2015), with more than 9 million patients, and the IMS Disease Analyzer Mediplus, with more than 2 million patients (Thomas et al. 2014).

## Adoption of Electronic Health Records in Different Countries

A number of organizations and researchers have attempted to estimate EHR adoption rates in recent years. However the use of different definitions for an EHR, stages of adoption and the degree of usage and functionality can vary greatly from one organization to another or even among or departments in a single organization. It may not be clear whether healthcare providers use specific EHR functions such as computerized physician order entry (CPOE) or decision support system, even if they report having an EHR fully deployed.

The Electronic Medical Record Adoption Model (EMRAM) of the Healthcare Information and Management Systems Society (HIMSS), a nonprofit organization aimed at improving quality, safety, cost-effectiveness, and accessibility in healthcare through the use of ICT and systems management, incorporates methodology and algorithms to automatically evaluate hospitals regarding accumulated capabilities of EMRs. It establishes adoption levels that allow comparisons between institutions. With this information, organizations can better plan the deployment process for complete electronic patient records to reach the full capacity of EHRs (Table 3.3).

**Table 3.3** Electronic Medical Record Adoption Model

| Stage | Cumulative Capabilities |
|---|---|
| 0 | The organization has not installed all of the three key ancillary department systems (laboratory, pharmacy, and radiology). |
| 1 | All three major ancillary clinical systems are installed (pharmacy, laboratory, and radiology). |
| 2 | Major ancillary clinical systems feed data to a clinical data repository (CDR) that provides physician access for reviewing all orders and results. The CDR contains a controlled medical vocabulary, and the clinical decision support/rules engine (CDS) for rudimentary conflict checking. Information from document imaging systems may be linked to the CDR at this stage. The hospital may be health information exchange (HIE) capable at this stage and can share whatever information it has in the CDR with other patient care stakeholders. |
| 3 | Nursing/clinical documentation (e.g., vital signs, flow sheets, nursing notes) is required and is implemented and integrated with the CDR for at least one inpatient service in the hospital; care plan charting is scored with extra points. The Electronic Medication Administration Record (EMAR) application is implemented. The first level of clinical decision support is implemented to conduct error checking with order entry (i.e., drug/drug, drug/food, drug/lab conflict checking normally found in the pharmacy information system). Medical image access from picture archive and communication systems (PACS) is available for access by physicians outside the Radiology department via the organization's intranet. |
| 4 | Computerized practitioner order entry (CPOE) for use by any clinician licensed to create orders is added to the nursing and CDR environment, along with the second level of clinical decision support capabilities related to evidence-based medicine protocols. If one inpatient service area has implemented CPOE with physicians entering orders and completed the previous stages, then this stage has been achieved. |
| 5 | The closed-loop medication administration with bar coded unit dose medications environment is fully implemented. The EMAR and bar coding or other auto identification technology, such as radio frequency identification (RFID), are implemented and integrated with CPOE and pharmacy to maximize point of care patient safety processes for medication administration. The "five rights" of medication administration are verified at the bedside with scanning of the bar code on the unit does medication and the patient ID. |

*Continued...*

**Table 3.3** Electronic Medical Record Adoption Model *Continued*

| Stage | Cumulative Capabilities |
|---|---|
| 6 | Full physician documentation with structured templates and discrete data is implemented for at least one inpatient care service area for progress notes, consult notes, discharge summaries or problem list & diagnosis list maintenance. Level three of clinical decision support provides guidance for all clinician activities related to protocols and outcomes in the form of variance and compliance alerts. A full complement of radiology PACS systems provides medical images to physicians via an intranet and displaces all film-based images. Cardiology PACS and document imaging are scored with extra points. |
| 7 | The hospital no longer uses paper charts to deliver and manage patient care and has a mixture of discrete data, document images, and medical images within its EMR environment. Data warehousing is being used to analyze patterns of clinical data to improve quality of care and patient safety and care delivery efficiency. Clinical information can be readily shared via standardized electronic transactions (i.e., CCD) with all entities that are authorized to treat the patient, or a HIE (i.e., other nonassociated hospitals, ambulatory clinics, subacute environments, employers, payers, and patients in a data sharing environment). The hospital demonstrates summary data continuity for all hospital services (e.g., inpatient, outpatient, ED, and with any owned or managed ambulatory clinics). |

*HIMSS Analytics (HIMSS, 2016c).*

However, to attend heterogeneity in rules for prescribing, validating, and distributing drugs for different regions, the original EMRAM was adapted and today there are different versions of the model for United States, Europe, and Asia-Pacific regions (HIMSS, 2016a—c).

## *United States and Canada*

In an attempt to create an EMR for most Americans by 2014, the US government established the Office of the National Coordinator for Health Information Technology in 2004 to promote and coordinate health information technology. Four goals were identified to guide the adoption of IT in the public and private healthcare sectors:

- Adoption of EHRs.
- Establishment of a secure national health information network.
- Use of personal medical records by individual patients.
- Use of research, dissemination of evidence, and quality measurement to improve the public health.

In the United States, in an attempt to create an EMR for most Americans, the US government established the Office of the National Coordinator for Health Information Technology in 2004. However, according to the last survey by the HIMSS, only 4.3% of hospitals are completely paperless. Similarly, in Canada, only 0.2% of hospitals are completely paperless, nearly

60% of the hospitals have only initial levels of EHR implementation and further efforts from government are being placed for EHR implementation (HIMSS, 2016c).

Although the progress seems to be slow, the results are considered significant in the light of the 19 billion dollars allocated by the Congress for the adoption of EHR and other health information technology. The major barriers to the implementation of EHRs among US hospitals that did not have EHR included: financial limitations (73%), maintenance costs (44%), cultural barriers (36%), uncertainty about return on investment (32%), and lack of IT training (30%). The study shows that physician resistance and inadequate capital are the major barriers for hospitals seeking to implement EHR (Jha et al., 2008).

### The European Union

The European Commission recommended to member states a set of guidelines for the deployment of interoperable EHR systems that facilitate cross-border exchange of patient health information. Developing such EHR systems should provide healthcare providers with a secure and timely access to the vital health information while protecting the patients' rights to confidentiality and privacy. The Recommendation facilitates ePrescription solutions through a set of guidelines for interoperability of emergency data, patient summaries, and medication record (European Commission, 2011).

The purpose of the guidelines is to make sure that EHRs systems in the European (EU) member states can interoperate (communicate to each other) to allow rapid access to vital patient information by healthcare providers across the EU. The objectives addressed by these guidelines include: establishing features of EHRs that allow exchange of vital patient information between systems; enabling share of health data; building network systems that cover all areas of healthcare, while meeting operational, legal, and training requirements.

The Commission aims to achieve and maintain cross-border interoperability of EHR systems by the end of the year 2015. According to the Commission, to achieve this goal member states are urged to undertake action at five levels: (1) the overall political leaders should make the necessary regulatory and financial environment to make eHealth infrastructure and services interoperable; (2) to create a common domain and interface that enable the national domains to interact; (3) to promote the use of technical standards and develop common interoperability platforms; (4) to agree on common priorities and specific applications; and (5) to improve education and awareness for monitoring and considering all intended and related developments.

However, according to the last survey by the HIMSS, which involved 1660 hospitals in Austria, Spain, Belgium, Denmark, Finland, France, Germany, Greece, Iceland, Ireland, Italy, Netherlands, Norway, Poland, Portugal, Slovenia, Spain, Switzerland, Turkey, and United Kingdom, only 0.2% of the hospitals are completely paperless, 41% are in intermediate stages, and 58% in initial levels of EHR implementation (HIMSS, 2016b).

### Asia Pacific

In the Asia-Pacific region the use of EHR systems has been increasingly expanded among healthcare institutions, and many regional/ national EHR projects have already been reported (Cho et al., 2010; Jian et al., 2007). However, a systematic review on the quality requirements of HER (Hoerbst and Ammenwerth, 2010) reported only few of deployment in Asia-Pacific region. Although China is still in a phase of changing paper records to electronic records, the government still realizes the importance to build up a national interoperable EHR system. According to the last survey by the HIMSS in Asia-Pacific region (HIMSS, 2016a) that involved 770 hospitals, only four hospitals in Asia are completely paperless (0.5%) with one of the hospitals located in South Korea and three in China. These hospitals do not use any paper charts, have a mixture of types of data in their EHR, and utilize data warehousing and systems to share clinical information. Thirty hospitals in Asia hold Stage 6 (3.9%). These hospitals are in Australia (1), India (6), China (9), Taiwan (4), Thailand (1), Singapore (8), and Malaysia (1).

### Developing Countries

Developing countries are a highly diverse group with different views and concerns (United Nations Statistics Devision, 2013). These are countries with low level of economy growth and inadequate technical and social infrastructures, especially in the area of ICT applied to healthcare.

In most developed countries the implementation of ICT across economic sectors has generally reached a high state of maturity. In developing countries, however, the implementation typically remains in the early stages, with different sectors of the economy at higher levels of maturity than others do. According to a WHO survey, 114 nations are currently working on National EHR systems (Webster, 2011). They have been implemented in HIV management in South Africa, Kenya, Rwanda, Ghana, Lesotho, Zimbabwe, Mozambique, Sierra Leone, Uganda, and Tanzania, as well as in various countries in Central and Latin America.

In Africa the major impetus for these efforts is the rise in infectious diseases as the leading cause of death (Hardcastle and Oteng, 2011), with HIV and multidrug-resistant tuberculosis as the most important. Patients with these conditions often require continuous treatment and long-term care, thus requiring an efficient record keeping system (Millard et al., 2012). Some chosen systems are "open source," characterized by open-source codes that are freely available to anyone who wishes to examine or change them for his or her own purpose. When compared to proprietary systems, open-source systems are much cheaper because, among other things, they are free of licensing and software upgrade costs and offer customisability for consumers (Webster, 2011). The possibility to use open-source solutions has provided more affordable systems such as the one implemented in 2012 at an urban emergency centre in Kumasi, Ghana (Forson et al., 2013).

The Brazilian National Health System (SUS) is a publicly funded healthcare system covering healthcare needs of more than 80% of the population of Brazil. The Brazilian Tripartite Commission (CTI), which represents the Ministry of Health, has framed standards for interoperability to be adopted by health information systems of the various levels of the SUS. The city of Sao Paulo adopted an open-source EHR system called SIGA Saude Health Information System, which annually serves around 14 million registered patients and stores data on 20 million patients from 702 health facilities (Webster, 2011). In Brazil the stage of maturity of ICT in healthcare is still low when compared to that of more developed countries, and poses great challenges to the government and stakeholders in the healthcare system. In this context, public policies must act to foster the adoption of several different technological solutions, which include improving both administrative management and clinical patient care. According to ICT in Health 2014 (CGI.br, 2014), in 29% of the hospitals, patient records are partially on paper and partially electronic. Electronic records were presented in 23% of the hospitals, with a 35% proportion for private hospitals and 9% for public hospitals. On the other hand, 45% of the hospitals keep their records totally on paper and, in this case, the proportion of public hospitals is 69%.

Improving the healthcare system requires the adoption and constant development of technological solutions that allow for the use of patient information and integrate management and clinical care processes delivered at different healthcare facilities. On the other hand, investment in training human resources is increasingly necessary to establish the required skills and competencies for extensive use of ICT by professionals in the sector.

## Challenges

A recent system review (Campanella et al., 2016) provides evidences that the use of EHR can improve the quality of healthcare by increasing time efficiency and guideline adherence, while reducing medication errors (David et al., 2013) and adverse drug effects (Pham et al., 2012). The number of benefits will grow as EHR use expands, costs fall, accessibility improves, and linkage to other relevant information from the individuals' ecosystem such as social, behavioral, environmental, genetic data, phenotyping, and clinical biobanks.

Developed countries have robust healthcare infrastructures that receive substantial funding and support from their governments. Besides, there is a strong support and motivation to accomplish goals associated with comprehensive development of successful IT systems, especially for EHR systems. These countries are able to make significant investments in research, development, and innovation in information systems that would meet the need of their particular healthcare system. On the other hand the healthcare infrastructure of many developing countries is still very limited. For many of these countries the delivery and management of healthcare services alone comes with many challenges (Braa et al., 2004). In many of these countries the development of healthcare IT solutions is faced with complex challenges such as inadequate funding, lack of human resources, and limited healthcare infrastructure.

Despite the benefits that EHR can provide, a proper implementation strategy is essential. An example of an effective strategy may be identified through the WHO (2006) guidelines for EHR in developing countries, which points six key actions:

- review the current health record system;
- try to emulate benchmark practices;
- involve the anticipated users of the system from the onset of discussions;
- train the users to the EHR system;
- evaluate the benefits of the implemented system;
- update the system when needed.

## HEALTH INFORMATION EXCHANGE

One of the advantages described in the 1990s for the adoption of electronic patient record was the possibility that all data from the clinical history of the patient were available to authorized persons anytime and anywhere, regardless of the country or information system he/she had attended (Ferrara and Sottile, 1999; Mori and Consorti, 1998; Iakovidis, 1998; Kohane et al., 1996).

Over the last 20 years, several groups have developed standards for exchange of information. In addition, several solutions were implemented that take into account aspects of security and data representation (König, 2005; Van Der Haak et al., 2003; Cao et al., 2003; Rassinoux et al., 2002; Smith and Eloff, 1999; Barber, 1998). Nevertheless the desired exchange of data between any institutions and systems has not been reached.

There are several reasons why the intended exchange of information was not reached yet. Some of them are exclusively regional, such as legal issues, lack of technical qualification, incomplete or faulty infrastructure, low priority, and low acceptance of use. Other problems are of economic-administrative nature, such as lack of investment or poorly targeted investments; incomplete/poorly defined legislation, lack of leadership for the adoption of standards, perception of low return of investment, missing global level agreements for the adoption of standards, and fear of legal actions relating to security of patient data.

Besides, there are also some technical problems that have not been completely solved: the multiplicity of existing standards with superposition between them; overly complex patterns; standards that do not meet the real needs of hospitals and/or health professionals; inadequate treatment of unstructured information (free text).

HIE is related to interoperability between systems. According to IEEE, [computer-based] interoperability is "the ability of two or more systems or components to exchange information and to use the information that has been exchanged" (IEEE, 1991). Interoperability can be divided into two levels: functional (or technical) and semantic (Gibbons et al., 2007). Functional interoperability is achieved at the level of messaging, i.e., the systems are capable of exchanging data properly (e.g., exchanging PDF documents using FTP), but they are not able to use the information computationally (e.g., to identify the patient's symptoms in a PDF report). Semantic interoperability allows processing the data computationally because there is an agreement on the semantics of the data exchanged between systems, i.e., the original meaning of information is preserved in the transfer between one system and the other.

To interoperate at the semantic level, it is necessary to develop a series of tools to support the process, from data entry that should be structured, going through the definition of vocabularies and structures representing the data (e.g., in case of representing a glucose measurement—which fields should be used? How one would define the reference values?). It is also necessary to define standards for exchanging messages. All these elements imply a major change in the way the activities are conducted within hospitals, making the exchange of information between health systems highly complex.

The efforts to meet these interoperability requirements created different standards for:

- *Vocabulary*. The Unified Medical Language System (UMLS) (US National Library of Medicine, 2009); Systematized Nomenclature of Medicine (SNOMED-CT) (International Health Terminology Standards Development Organisation, 2002); Logical Observation Identifiers

Names and Codes (LOINC) (Regenstrief Institute, 1994); Medical Subject Headings (MESH) (US National Library of Medicine, 1999); International Coding Diseases (ICD) (WHO, 1992); among others.

- *Exchange of information.* Health Level 7 (HL7 1989), Digital Imaging and Communications in Medicine (DICOM) (NEMA, 1998), Cross Enterprise Document Sharing (XDS) (IHE, 2006), Integrating the Healthcare Enterprise (IHE) (IHE, 1998), CEN ISO/IEEE 11073 Health informatics—Medical/health device communication standards (ISO et al., 2004), Health informatics—Electronic Health Record communication—EN 13606 (EN 13606 Association, 2010), the outdated CORBA security services for health information systems (CORBAmed) (OMG, 1997), among others.
- *Structuring of information.* Health Level 7 version 3 (Health Level 7 2015), Clinical Document Architecture (CDA) (HL7 2007), Continuity of Care Document (CCD) (ASTM & HL7 2007), openEHR archetypes (openEHR, 2006). There are also important groups responsible for creating new standards, such as the International Organization for Standardization's Technical Committee on Health Informatics (ISO/TC 215) that develops standards for all data interoperability levels (ISO, 1998) and a similar group, the European Committee for Standardization Technical Committee (CEN / TC 251), which also develops several standards for interoperability (CEN, 1993).

There are other categories of health data exchange, besides the exchange between EHRs that must be mentioned:

- *Governmental demands for information gathering, in order to control epidemics and adverse events*: Government demands (at municipal, state, and federal levels) reflect the need for monitoring the population for determining diseases outbreaks, population growth and estimation of prevalence of diseases. This exchange of health data is regulated by governments and can be performed through the use of health standards (such as HL7, ICD-10, and openEHR), computational standards (Web Services, SOAP, XML), standards developed by the government (Jian et al., 2007; Macedo Pinho Rotzsch et al., 2005) or a mixture of those. The limitations of government initiatives are related to several factors, such as the different levels of informatization of healthcare facilities, which may vary depending on the country, ranging from fully computerized to fully paper-based; the implementation at national level depends on physical infrastructure (computers, network, and bandwidth), clear implementation guidelines and technical training; the quality of the information collected depends on the employees (e.g., experience, training) and the computational implementations (some possible limitations are: bad interpretation of the standards, incompatibility with existing information systems, faulty programming, different versions of software and the use of different versions of the standards); governmental initiatives also depend on their managers, who must understand the health sector and implement solutions compatible with the level of development of the country; appropriate solutions depend on the existence of healthcare Institutions (representatives of society) that can influence the definition of governmental health policies allowing for greater dialogue with society and adoption by health entities. It is also essential the continuous investment in training and maintenance of the necessary infrastructure of private and public healthcare.

■ *Multiinstitutional research projects.* Multicentric research projects have a clearly defined common goal among the participants, which facilitates the definition of the data to be exchanged. In addition, special attention is given to correction of the shared data, which ensures the quality of information. A multiinstitutional research uses nonstandard tools (in the sense that are defined for each project), such as spreadsheets or internet-based tools (such as I2B2 (Murphy et al., 2006), Redcap (Harris et al., 2009), and XNAT (Marcus et al., 2007)). Currently there is great interest in multicentric trials stimulated by funding agencies, which has led to greater concern with the standardization and automation of the exchange of information between the institutions, leading to creation of specialized software for clinical research.

■ *Systems for health monitoring by the own patient.* The increased availability of Internet and the higher number of mobile devices (laptops, tablets, and smartphones, some of which contain more sophisticated sensors, such as heart rate monitor) lead to a growth in the number of programs for monitoring personal's health. These initiatives are usually supported by society and government, since the aging of population has increased the incidence of chronic diseases (such as high blood pressure, diabetes, and cardiac diseases), whose treatment is expensive and its consequences can be minimized with the (preferably early) adoption of healthy habits, balanced diet, and exercise (Paré et al., 2007). These programs can range from PHR to systems for controlling the amount of water intake throughout the day, including fitness programs, diabetes control, localization of hospitals and health centres, among others. In general, these programs are web-based and do not follow health standards for storage and exchange of data, but provide the benefit of being used to collect data (e.g., programs for diabetes control) and being easily accessed during medical appointments. These applications meet the needs of the user, who gains control over his own health data, a characteristic that is often absent in EHRs. The user can also show his medical data to different doctors (e.g., using programs like HealthVault (Microsoft, 2016)). A more detailed discussion about the uses of mobile devices in Healthcare is presented in "Mobile Devices" section of this chapter.

- *Informal consultations among health professionals.* The higher availability of Internet gave opportunity to the informal consultations among healthcare professionals, who use messaging systems (such as WhatsApp and Skype), as well as e-mail, social networks (such as Facebook and Google + ), and specialized applications (such as OsiriX (Pixmeo SARL, 2004) and ResolutionMD (Calgary Scientific, 2013)) for patient care and information exchange. These initiatives are difficult to be accounted for, because they often do not guarantee the security of patient data (e.g., not being Health Insurance Portability and Accountability Act (HIPAA) compliant) and are, in many countries, prohibited (Boulos et al., 2014). These initiatives may indicate the lack of easy and practical programs and also a lack of understanding of the risk of exposing patient data. In some countries, where the primary access to Internet is done by mobile phones, those applications end up being viable alternatives (Hampshire et al., 2015; Nsanzimana et al., 2012). Currently, there are some HIPAA (US Department of Labor, 2015) compliant applications that propose to be a reliable alternative to exchange medical information, such as DocBookMD (J&H MEDSOFT, 2016) and Lua Technologies (2016). The exchange of data in most of these systems is not standardized yet. In "Social Media" section we discuss more deeply the uses and risks of the healthcare communication using social media.

## Challenges

HIE adoption has several layers of challenges:

- *Technological issues.* (1) Harmonization of the existing standards and (2) overly complex standards.
- *Legal definition.* (1) Mandatory use of standards for information exchange; (2) exchange of data between different countries; (3) HIPAA-like rules for patient confidentiality and privacy (outside the United States); (4) storage of patient information outside the country (e.g., cloud); and (4) use of "apps" for healthcare.
- *Economic-administrative issues.* (1) Investments in infrastructure; (2) investments in training; (3) set HIE as a priority for healthcare; (4) economic sustainability of HIE initiatives; and (5) informatization of healthcare facilities in developing countries.

Much of the necessary efforts to face these challenges are already being addressed, but the progress is still very slow. The HIE is a desire of healthcare consumers, but it still not the major priority. The growing availability of access to digital information will increase the demand for access to EHR and related patient data making necessary a faster adoption of solutions for interchange.

## TELEMEDICINE

To write about telemedicine is a complex task, since the concept of the telemedicine encompasses a wide variety of activities. World Health Organization (WHO) defines telemedicine (Telematics, 1998) as "the delivery of healthcare services, where distance is a critical factor, by all healthcare professionals using information and communications technologies for the exchange of valid information for diagnosis, treatment and prevention of disease and injuries, research and evaluation, and for the continuing education of healthcare providers, all in the interests of advancing the health of individuals and their communities ". This is a broad definition that encompasses several "tele" concepts as teleeducation in medicine and teleradiology. It is important to note that this definition is not perfect, since it does not restrict the manner in which data is exchanged (e.g., an exchange of e-mails can be considered telemedicine?), nor mention the safety aspects for the exchange of data (as encryption and authentication of the participants).

The common point among the different telemedicine definitions is that the participants must be separated geographically and some form of IT is used to exchange data. Telemedicine can occur between two health professionals (e.g., to get a second opinion), between a patient and a health professional (teleconsultation), or between different health professionals as in the discussion of clinical cases and teaching medicine (teleeducation and teleclass). Other forms of telemedicine are performing remote diagnostics (remote diagnostics), remote monitoring (telemonitoring) and performing surgeries remotely (telesurgery). The type of interaction between the participants is done through electronic media such as video, image, audio, chat, whiteboard, desktop sharing, e-mail, SMS, among others.

The communication between the participants can be synchronous or asynchronous. In the synchronous communication the exchange of messages is immediate or with little delay, as in video conversations. In asynchronous communication, message exchange occurs with a variable time interval, such as the exchange of emails. The asynchronous communication generally demands less computational resources (lower bandwidth, network quality lower and lower processing capacity) while synchronously requires more technology and infrastructure.

Telemedicine is described in the literature as a potential source of several benefits, including:

- improvement in the quality of medical care in remote locations (such as rural areas) or difficult to access (Kolbe et al., 2015);
- possibility to discuss cases in a collaborative network (Coppola et al., 2016);
- improved quality of care in developing countries, where there is a shortage or a nonhomogeneous distribution of qualified professionals (Kashif Mirza et al., 2008; Pagni et al., 2011);
- service outsourcing, e.g., the writing of reports remotely (Dixon and FitzGerald, 2008; Wong et al., 2005; Kangarloo et al., 2000) or with the use of Intensive Care Units surveillance centers (Kumar et al., 2013a), reducing costs;
- improving patient education and empowering patients to manage their own health (Paré et al., 2007).

Reducing costs through telemedicine use has not been proven yet, since the results published in the literature are very heterogeneous and use different calculation methodologies (de la Torre-Díez et al., 2015; Ward et al., 2014; Kumar et al., 2013a). These problems are partly due to the variety of activities that comprise telemedicine, which hampers a general comparison between telemedicine solutions. According to the Global Observatory for eHealth series report (Ryu, 2012), teleradiology is the modality with the highest proportion of services established in the world (33% in 2010), followed by telepathology (17%), teledermatology (16%), and telepsychiatry (13%).

Developing countries are some of the major beneficiaries of Telemedicine. These countries have several limitations (financial, infrastructure, training, distance) that make Telemedicine a good candidate to improve local healthcare services. For this reason, there are various initiatives of nongovernmental organizations and humanitarian groups for deployment of Telemedicine solutions to low-income countries. One of the oldest initiative is the Swinfen Charitable Trust, which offers a free second opinion service to developing countries since 1998 (Carr and Wootton, 2013). Volunteers answer queries using an asynchronous Web system (Store-And-Forward, SAF), which works as follows: (1) a new case is submitted by the referent (doctor or healthcare professional) that may include images; (2) next, a coordinator analyzes the submission, checking the quality and relevance of data and redirects the request to a chosen specialist; (3) the system notifies the specialist and awaits for an answer. If there is no specialist answer in a maximum of 48 hours, the coordinator redirects the case to another expert; (4) the specialist answers the question and the system notifies the referent that the answer is available; and

(5) the referent reads the answer. The technology used is simple and requires only Internet access and a Web browser, which can be the reason for the system's longevity. On the other hand the system is apparently underused, since it has been used at an average of 20 medical cases per month (in 2900 consultations until 2012). The causes for this low utilization are discussed by Wootton (2008), which cites as possibilities: (1) depreciation of the service for being free; (2) cultural problems preventing asking for help; (3) not choosing appropriate specialists; (4) lack of time in the referent activities to request for help; and (5) reluctance to ask for assistance to a foreign country. Similar numbers (16 cases per month) were obtained by an equivalent system, operated by Médicins Sans Frontières (Halton et al., 2014). Saliba et al. (2012) cite the lack of trust between professionals from different countries as a possible cause.

Developing countries receive international support from several humanitarian telehealth networks, which include (Scott and Mars, 2015): Réseau en Afrique Francophone pour la Télémédecine (RAFT) Network; PAN African e-Network; Swinfen Charitable Trust; Africa Teledermatology Project; Institute of Tropical Medicine (TM) Telemedicine; Pacific Island Health Care (PIHC) Project; Partners Online Specialty Consultations; Balkans e-Health Network; and Teletrauma.

There are several factors that can hinder the implementation of Telemedicine (Weinstein et al., 2014; Saliba et al., 2012):

- *Legal factors*. The laws of the country must determine: the mandatory documentation for Telemedicine; the quality assurance of services (e.g., audit, certification, and hospital credentialing); the need of informed consent; determination of liability for specialist and referent in case of negligence; recognition of the role of health workers in another country or state (work permit); the appropriate jurisdiction for participants from other countries; to maintain confidentiality and security of patient data; laws regulating telemedicine service reimbursement.
- *Sustainability factors*. As stated earlier the cost-effectiveness of telemedicine was not proved yet. There are also many pilot programs that did not go beyond their original funding, indicating that sustainability is an important issue. Financing can be provided by the government (regionally (Marcolino et al., 2013), country-wide (Sutton, 2011), or in international collaborative projects (Köhler et al., 2005)), nonprofit (Halton et al., 2014), private investment (Muehlberg et al., 2015), or a mixture of those (Bediang et al., 2014). Another important issue is the integration of telemedicine services within mainstream health systems to achieve scalability and sustainability. When applicable, it is necessary to create a business

plan for sustainability that could invest on (1) gap service coverage; (2) urgent service coverage; (3) mandate services; and (4) video-enabled multisite group chart rounds, for example (Masi et al., 2012).

- *Cultural and language factors.* The difference in language can lead to misinterpretation of situations causing difficulties in meetings or erroneous decisions. Cultural issues can cause lack of confidence in the opinion of a foreigner.
- *Contextual factors.* Endemic diseases may be unknown to the remote expert, leading to misconduct. Certain laboratory/imaging tests may be incompatible with local conditions (too costly or unavailable). There may be infrastructure problems, such as lack of electricity and limited network availability, which affect telemedicine services. There may be resistance to change and to the adoption of new technologies.

Another important topic for telemedicine is to provide healthcare access to rural areas. There are well-succeeded initiatives to attend rural areas both in developed and in development countries (Hess et al., 2005; Mariño et al., 2016; Qin et al., 2013; Duchesne et al., 2008; Versleijen et al., 2015; Kolbe et al., 2015). The reason why there are many successful deployments may be because rural areas are better controlled environment, where it is possible to adapt the local routine and the developed technologies to better serve the specific local. On the other side, generalized approaches (such as at national level initiatives) may not be suited for particular settings.

## Challenges

Among the several challenges for telemedicine, we can mention:

- *Legal.* (1) Reimbursement by services; (2) liability of the participants, in case of negligence; (3) mandatory informed consent; (4) minimal required documentation; (5) audit, certification, and credentialing.
- *Sustainability.* (1) There is still no proof that telemedicine is cost effective; (2) Developing countries need governmental assistance to maintain telemedicine programs.
- *Culture/language.* Solutions must cope with different cultures and languages.
- *Regional factors.* Solutions must be able to deal with limitations such as shortage of energy, limited training, high turnover, endemic diseases, unavailability of certain medical procedures (e.g., troponin results or cardiac ultrasound), among others.

The sustainability issue seems to be the most important because it will direct the investments on telemedicine and allow the development of real and durable solutions. We believe the legal problems will be solved when telemedicine is perceived by the population as an important asset for the quality of healthcare delivery. The use of smartphone applicative for healthcare, e.g., is requiring the positioning of government and health societies creating a common understanding on the subject. Cultural and language issues can be solved with proper training, and regional factors can be softened by developing robust applications and frameworks.

## MOBILE DEVICES

The fast global advances in wireless communications allied with the downsizing of hardware components, the development of faster CPU processors, and the increase in storage capacity allowed the creation of devices with the computing power of a desktop inside the small case of a mobile phone, the smartphones. The emergence of this new technology opened a wide range of new services that changed our lives in deep ways. The communication using Internet, smartphones and, more recently, social media is changing the way we interact with each other as well as the way services are provided by the several industries.

Healthcare is not exception in this scenario and the coupling of different technologies applied to medical area allows, for instance, the acquisition and transmission of medical images through mobile devices. This sophisticated application is part of a wide spectrum of the so-called mobile health applications, or mHealth apps. Actually the spread of mobile health initiatives around the world in the last decade represented a striking movement of the healthcare sector.

There is not a single definition mobile health, or mHealth. The WHO described mHealth or mobile health as "medical and public health practice supported by mobile devices, such as mobile phones, patient monitoring devices, personal digital assistants (PDAs), and other wireless devices" (World Health Organization, 2011).

The rapid development of healthcare applications based on mobile platforms was driven mainly by the increasing ownership of smartphones around the world. In addition to the pervasiveness of this equipment, other attributes of mobile technology help to explain the high number of mHealth solutions. They are: (i) low cost when compared to fixed lines; (ii) simpler use when compared to other Internet interfaces; and (iii) the possibility to connect with sensors to gather data relevant to healthcare assessment and delivery (Akter and Ray, 2010; Varshney, 2014; Baig et al., 2015).

mHealth has the potential to change the way healthcare is delivered by introducing new channels and tools to the healthcare professionals and the patients. The existing definitions of mHealth encompass a wide spectrum of services and applications, from calories counters to international collaboration projects aiming to provide complex homecare services, like remotely monitoring a renal care system attached to the patient (Simonis et al., 2011).

In the next sections, we will present some examples of mHealth applications, for both single downloadable apps—or consumer mHealth—and university or governmental projects.

## mHealth Projects

Some important goals of mHealth projects are: to provide faster and more correct access to healthcare and patient information, by improving healthcare processes; to allow prevention, remote monitoring and assessment of chronic patients; to help in emergency processes by speeding detection and transport and getting information about patients while providing first aid (Varshney, 2014). To achieve some of these goals, mHealth use precise sensors to monitor patient or environment variables (Kumar et al., 2013b).

There are mHealth projects in the whole world, though the main purpose of these initiatives varies accordingly to the region (Crul, 2014; Varshney, 2014). In many developing countries the focus is to improve the access to basic healthcare, by allowing remote prevention and monitoring unserved or underserved communities. For such areas, mHealth can be a fundamental aspect of healthcare delivery (Varshney, 2014) and wireless technologies help us to face endemic and epidemic diseases (Iwaya et al., 2013). On the other side, in countries with a wide availability of healthcare services, mHealth has a supportive role and is mainly oriented to reduce costs, to engage patients in their own care, and to deliver healthcare with increasing quality (Crul, 2014). However, one should not underestimate the value of the potential role of mHealth in developed regions. Due to the aging of population and the rising of healthcare costs, the development and adoption of mHealth solutions may represent an important aspect for the maintenance of high level of healthcare delivery of those regions.

In the remaining of this section, we present some example of projects carried out in developing and developed regions. The number of projects mentioned is very small given the many mHealth projects developed and under development all over the world. The purpose here is to present example applications that may help to settle important healthcare issues of these regions.

### mHealth Projects in Developing Countries

Several areas of application have been described in the literature for low- and middle-income countries, including (Jahan and Mozammel Hoque Chowdhury, 2014):

- emergency systems;
- coordination, management, and supervision of healthcare workers;

- decision support Telemedicine to remote clinicians;
- monitoring of remote patients;
- management of health services;
- training general populations on relevant health issues;
- continuous training of healthcare workers;
- involvement of communities in health promotion.

Targeting at some of these goals, developing countries have witnessed a number of initiatives of mHealth to help the work of frontline health workers. These professionals, who provide the first aid for people living in remote and rural areas, are midwives, nurses, pharmacists, doctors and community health workers (CHWs). Most of the applications aim particularly to improve the actions of CHWs, who are members of a community that helps to provide basic healthcare to their own community. The existence of those professionals helps in the shortage of highly educated healthcare professionals in some regions of low- and middle-income countries (Agarwal et al., 2015). There are many aspects that can be improved in the fieldwork of these professionals, such as poor data collections tools, lack of parameters to prioritize household visits, lack of training and reference material, and improvement of communication with other professionals in potential troublesome situations. Many published works evidence the effectivity of mHealth initiatives involving CHWs in spite of their level of education or training (Agarwal et al., 2015).

Many applications are related to the use of mobile phones to collect healthcare data. Pilot projects using this technology were launched in many countries. An evaluation of the effectiveness of collecting health information from primary care to detect disease outbreaks was performed in India, with a conclusion that the data submission using mobile phone was a cheaper option than using paper tools (Ganeshan et al., 2011). Similar experiences of implementing support of primary healthcare at communities using mobile phones were performed in Ghana (Macleod et al., 2012) and Bangladesh (Alam et al., 2010). In South Africa, there is a work reporting the record and submission of data from multidrug-resistant tuberculosis patients during the intensive therapy phase (Chaiyachati et al., 2013). A description of improving antenatal care and reducing child mortality using a mobile application was reported in Thailand (Kaewkungwal et al., 2010). In Brazil, there was a project named Borboleta for collection and visualization of health-related data including images and video (Duarte et al., 2010), and a recent work presented the deployment of a secure mHealth system applied to support primary healthcare communities on a large scale (Sa et al., 2016). Comprehensive surveys of mHealth initiatives have been published in India (Ganapathy and Ravindra, 2008), China (University of Cambridge and China Mobile, 2011), and Brazil (Iwaya et al., 2013).

A main problem reported from most of these works is that the implementations described are part of pilot projects that do not succeed to become regional or national scaled implementations (Agarwal et al., 2015; Iwaya et al., 2013). Government and donors had even coined a term—"pilotitis"—for this inability of scaling, i.e., characteristic of most mHealth projects (Franz-Vasdeki et al., 2015). Another weakness lies in the lack of evidence about the possible improvement on health system efficiency that may lead to better healthcare delivery to the population (Agarwal et al., 2015).

## mHealth Projects in Developed Countries

The major driving force for the launching of mHealth applications in developed countries has been the increasing of costs to delivery healthcare for an aging population and their associated chronic diseases. Among the most prevalent diseases are cancer, type 2 diabetes mellitus (T2DM), chronic obstructive pulmonary disorder (COPD), cardiovascular diseases (CVD), and psychiatric disorders (PWC & GSMA Connected Living Programme, 2013). According to WHO data from 2012, this set of disease are the major causes of death in high-income countries World Health Organization, 2012. There are evidences that mHealth may improve the healthcare delivery for chronic diseases in aging societies once it can deliver solutions for self, assisted, and supervised healthcare (Chiarini et al., 2013). mHealth technologies also can help to remotely monitoring the patients using sensors attached to the patient (Varshney, 2014). On the other side, some of these chronic conditions are largely preventable, and mobile health can be of great help by providing health promotion tools (exercises, diets, stop smoking).

In this context, some important applications of mHealth are in the following tracks (PWC & GSMA Connected Living Programme, 2013):

- Wellness and prevention—mHealth apps can provide information that influences patients and general population to improve lifestyles with healthier food habits, exercising, and help in quit smoking;
- Accelerating diagnosis—promoting early detection of diseases thus helping to decrease the severity of the symptoms;
- Decrease hospitalization—by remotely treating and monitoring the patients, and still allowing interventions when required;
- Optimize healthcare delivery—by providing mobility to healthcare personnel when analyzing patient data.

The European Union (EU) has funded several programs for monitoring and diagnosing patients remotely by using mHealth and wearable technologies. Applications targeted diverse conditions, including: (1) Chronic Obstructive Pulmonary Disease and Chronic Kidney Disease: project "Chronius" (Bellos et al., 2011); (2) depression: "ICT4Depression" (Warmerdam et al., 2012) and "Pshyche" (Javelot et al., 2014); (3) cardiovascular diseases: "Bravehealth" (Dhukaram et al., 2011); (4) diabetes: "Reaction" (Warmerdam et al. 2012; Manousos et al., 2013). A more challenging project, "Nephron + ," developed a portable artificial kidney device that can be remotely monitored by the patient's smartphone (Simonis et al., 2011).

In the United States, "Mobilyze" and "Medlink" were projects funded by the National Institute of Health for treatment and support of patients suffering from depression (Ben-Zeev et al., 2015). A third project aiming at a high risks patients suffering from schizophrenia, named "FOCUS," was funded by the Centers for Medicare and Medicaid Services (Ben-Zeev et al., 2015). mHealth projects for glucose monitoring that also aimed at behavioral interventions of adolescents with type I diabetes were developed in Canada (Cafazzo et al., 2012) and United States (Carroll et al., 2011).

As in the case of projects of developing regions, many of these projects remained as pilots.

## Consumer mHealth

The group of application discussed so far is usually the result of big projects involving research groups, hospitals, or government teams. These projects usually involve complex applications, which try to integrate data from the applications with formal healthcare systems, like hospitals EHRs or government healthcare data. On the other hand, a large group of mHealth applications developed with commercial purposes grows at an amazing pace. Data from 2015 indicated more than 100,000 consumer mHealth applications for iOS and Android platforms (Baig et al., 2015).

A large group of consumer apps intends to manage chronic diseases. Applications for self-management of diabetes usually use diaries and logs with the user feeding the data, and the recent study in the United Kingdom showed that the patients suffering from this condition want intelligent, personalized, and contextually aware advice (Katz et al., 2016). The authors point out that future apps shall automate connected sensors and use intelligent processing and visualization to assist effectively the patients. Another assessment on diabetes apps showed that the insulin dose calculators usually present on these software may lead to incorrect dose recommendation exposing patients to risks (Huckvale et al., 2015). For evaluating hypertension management apps, Kumar et al. (2015) analyzed 107 apps, most of which were designed for health management by tracking blood pressure, weight, or body mass index, medication adherence and providing information about the condition and diet. An important finding of the study, however, was that most of the apps lacked clinical validation. A study from Widmer et al.

(2014) provided the evidence that a smartphone-based application intended to deliver the Mayo Clinic's cardiac rehabilitation program was an effective preventive strategy. A study from the Imperial College evaluated 185 apps related to breast diseases (Mobasheri et al., 2014). Although the study evidenced some pros of such applications by providing relevant information to the patients and allowing the management of appointments and exams, it also showed a disturbing lack of evidence base and medical professional involvement in their development, which raises safety concerns. A massive group of apps is related to physical activities, and a study performed by Conroy, Yang and Maher with 167 top-ranked apps related the absence in these apps of behavior changing techniques, which are considered important among experts (Conroy et al., 2014). A more comprehensive study was dedicated to evaluate the apps for the eight most prevalent health conditions according to the WHO: iron-deficiency anemia, hearing loss, migraine, low vision, asthma, diabetes mellitus, osteoarthritis, and unipolar depressive disorders (Martínez-Pérez et al., 2013). The results have shown that most of the apps are dedicated to diabetes and depression while other conditions have less attention.

It has to be pointed out that the quality of these applications is still an open issue, and quality control mechanisms are necessary to guarantee patient's safety and privacy (Fernandez-Luque and Bau, 2015). A particular critical point, as many authors emphasize, is that most of the apps lack clinical validation.

A very interesting initiative of the European Commission is the *European Directory of Health Apps* published since 2012, which presents information about health apps to help the patients on a large number of medical conditions. In the 2015–16 edition there are 300 apps, all recommended and rated by patient groups (PatientView, 2016).

## Challenges

mHealth initiatives have been launched all around the world, in developed and developing regions, for more than a decade now. Many of these initiatives were very good projects, which met the goal to demonstrate the potential use of this technology. However the great majority of them remained as pilot projects and never were fully deployed for use in large scale (Labrique, 2013; Tomlinson et al., 2013; Iwaya et al., 2013). There are many challenges that must be faced to move from the project to the full deployment, and in the following we are going to discuss some of them.

- *Security*. Security and privacy of patient data are sensitive issues in any healthcare application. General EHealth quality controls and mechanisms to guarantee patient's data availability, confidentiality, integrity, and authenticity also apply to mobile applications. Furthermore, some specific needs add to the list, such as reliable data transmission and timely data delivery. A strong debate is now taking place about the need to regulate mHealth applications. On the other side, there are some voices warning about a possible danger of overregulation (Fernandez-Luque and Bau, 2015). In the United States the Food and Drug Administration (FDA) has published a guidance document for informing about regulatory rules for mobile platform software applications (Food and Drug Administration/US Department of Health and Human, 2015). The FDA guidance does not address data privacy protection, which is performed by the HIPAA in the cases where it is pertinent (Petersen and DeMuro, 2015). Europe is on the final phase of approving a Code of Conduct on privacy for mHealth apps, with the aim to increase the compliance for developers at European level (European Commission, 2016).

- *Integration to national and international healthcare programs and standards.* The ultimate goal of mobile applications is to become part of the healthcare systems, in such a way that their data can be distributed across different health applications, and connect to electronic healthcare systems and national or regional healthcare programs (Chen et al., 2012). For achieving the goal, the solutions will have to adopt not only the above-described security rules, but also to enforce national and international standards for data, information and procedures related to health sector (Ballesteros and Mayas, 2015) as discussed in "Health Information Exchange" section. Patients must also be included in the discussion of the type of tools that would be beneficial for helping them in monitoring and assessing their health conditions. National (via Ministries of Health) and international health organizations will have to be involved in the development of mHealth solutions, providing the necessary guidance for those who are developing, so the new tools can be targeted at relevant priorities. Groups and developers will also have to adopt clinical guidelines and conduct high-quality controlled trials before the delivery of products.

- *Metric for quantify the real benefits of mHealth.* Although most, if not all, of the woks describing mHealth initiatives claim the potential of the technology to improve the coverage, the quality and the effectiveness of healthcare systems, offering a wide spectrum of benefits for patients. However, robust evidences supporting that claim are still lacking (Galliber et al., 2008; Prgomet et al., 2009; Tomlinson et al., 2013) and a major gap of the mHealth projects is that their alleged benefits and long-term results are not precisely measured and evaluated (Aranda-Jan et al., 2014). A recent review on the theme by Mookherji et al. (2015)

showed that indeed there is a lack of evidence to support the assumption of mHealth benefits for health service, utilization, behavior, and outcomes. An interesting result of this review was that most of the times, the evidence base is in the field of computer science and not on health impact. This result can be explained by the fact that health deliver practitioners do not play a key role at mHealth field, a position that belong to people from technology sector (Franz-Vasdeki et al., 2015).

- *Sustainability.* A factor of great relevance to "pilotitis syndrome" of mHealth initiatives is the lack of financial sustainability after the project phase finishes. Generally the project managers do not perform cost-effectiveness of deployment and maintenance of large-scale projects (Leon et al., 2012). The absence of this analysis, allied to lack of evidence base of the results, can prevent investments from government and industry partners in scaled initiatives (Mechael et al., 2010). Finding sustainable and cost-effective models that can be integrated to established care-financing systems would be a driving factor to turn projects into scaled system deployments.

Thus, in a near future, health and economic indicators of quality and effectiveness should be discussed and adopted, so the impact of mHealth to improve health outcomes can be measured.

## SOCIAL MEDIA

Social media concept followed the concept of social network sites, whose classic definition was given by Boyd and Ellison (2008)

> We define social network sites as web-based services that allow individuals to (1) construct a public or semi-public profile within a bounded system, (2) articulate a list of other users with whom they share a connection, and (3) view and traverse their list of connections and those made by others within the system. The nature and nomenclature of these connections may vary from site to site.

In a short period, social media sites became the most popular internet destinations, allowing people to communicate without the limits of geography, time, and cost (Drazen et al., 2012). Today there are many definitions of social media and no one is definitive. Merriam-Webster gives a simple and broad definition that captures the main concepts of the term:

> Social Media are forms of electronic communication (as Web sites for social networking and microblogging) through which users create online communities to share information, ideas, personal messages, and other content (as videos).

Another major worldwide trend is the use of Internet as an important source of information for people suffering from any disease, as an attempt to understand better the condition itself and to find treatments that are more effective. Today, it is quite common that patients arrive to medical consultations with self-diagnosis from online research (Mannu, 2015). Furthermore, individuals want to publish content about health topics and, in case of diseases, exchange information and share experiences with others suffering from the same condition. The social media sites appear as the natural tools for these activities, and in fact, patients are using them extensively for these purposes. In addition to patients, healthcare professionals, healthcare institutions and government healthcare agencies are using social media to share information with the population.

In the following section, we present some general view of the use of social medial tools in medicine and healthcare by these groups. This view is not intended to be extensive, and for a wider picture of how many of these tools are being used in healthcare refer to a review by Grajales et al. (2014). The use of these tools in healthcare applications raises some problems, which will be discussed in Section Challenges.

## Healthcare Social Media Uses

### Patients and Healthcare Professionals

The healthcare information exchange between patients and between healthcare professionals is a natural outcome of the ubiquitous use of social media tools. There is a large number of works describing the use of social media in healthcare, the majority of which related to discussion forums (Hamm et al., 2013). Indeed, many studies show the connection power these tools can provide to patients, supporters, and caregivers (Kahn, 2008; Moorhead et al., 2013; Frost and Massagli, 2008).

Patients use different types of social media to create online communities, in which they can share experiences and obtain and/or give emotional support. There are platforms intended to congregate patients with different diseases, as PatientsLikeMe (that actually congregates several communities for each disease), MedHelp, DailyStrength, and CureTogether.

Online support communities have proven to be particularly relevant for patients with chronic or rare diseases (Zaidan et al., 2015; Bender et al., 2011; Frost and Massagli, 2008), and there is a huge number of platforms dedicated to social media networks of specific health conditions, namely chronic ills. Social media sites sharing information about several cancer types (IHadCancer, Esperity and KnowCancer, Melanoma Patient Network Europe) and diabetes (TuDiabetes, TypeOneNation, DiabetesSisters, and Diabetes Hand Foundation) are the most cited. They all provide supportive communities to patients and their families, where they can receive information about new treatments and insights, and receive support from other persons with the same condition.

Healthcare professionals also have a wide agenda of social media use, including: mining and sharing of information; communication with peers, trainees, or supervisors; service advertisement. Some of them use social media to interact with patients. This last group is still the minority, but is increasing rapidly (Chretien and Kind, 2013). There are physicians communities for exchanging knowledge on clinical facts, drugs and difficult cases (Sermo, Doximity, QuantiaMD, Osmosis, iMedExchange), nurses communities (NursingLink, AllNurses, NurseConnect), Physiotherapists (Connectin Physios), among many others.

An interesting outcome of a survey conducted in 2013 by Antheunis et al. (2013) is that, as a rule, the groups of patients and caregivers use social networks for different purposes. The contact with other patients was found as the main reason for patients to use social media; while marketing and contact other professionals were the primary reasons of healthcare professionals. This result agrees with the general stablished use of networking sites, which is to meet people with similar interests (Bender et al., 2011). Moreover, although the use of social media to facilitate the communication between patients and healthcare professionals appears to be straightforward, this channel presents important challenges related to data privacy and confidentiality, boundaries in the relationship between the patient and the professionals, and professional behavior (Ibarra-Yruegas et al., 2015).

### Healthcare Institutions, Health Insurance Companies, Pharmaceutical Companies, Government and International Agencies

The adoption of social media as a communication channel happened more slowly in healthcare than other areas (Drazen et al., 2012). Eventually, however, healthcare institutions have realized the importance of this new media and today the traditional mail and telephone communication between hospitals and their patients outside clinical settings are being replaced to channels like social media and smartphone apps. Numbers from 2014 show that virtually all US hospitals are part of at least one social media: 94% have a Facebook page, 99.14% have a Yelp page, and 99.41% have check-ins on Foursquare (Griffis et al., 2014).

Governmental healthcare campaigns and policies also use social media and smartphones as a communication channel with the population. Health Ministries of countries all over the world have channels on Facebook, YouTube, LinkedIn or Twitter, and communicate with their citizens by posts, tweets, and podcasts. International healthcare organizations, such as The

World health Organization (WHO), Médecins Sans Frontières, and Red Cross use social media to communicate with general public, donors and those possibly in need. Finally, insurance and pharmaceutical companies also have media social as an important communication channel with their clients.

### Whatsapp

In many parts of the world the instant messaging platform WhatsApp has become a very popular communication tool in care-related applications. The possibility of sending pictures, audio, videos, and archives related to a particular question and receive immediate answers are compelling features that made the tool so popular among doctors and patients in many countries (Thota and Divatia, 2015; Alam et al., 2010; Giordano et al., 2015; Petruzzi and De Benedittis, 2016).

A survey with 1084 doctors from eight countries (Brazil, China, France, Germany, Italy, Spain, United Kingdom, and United States) showed that 87% of respondent doctors in Brazil, 61% in Italy and 50% in China reported using the platform. On the opposite side, only 2% of the UK doctors and 4% of US doctors reported using it (Mannu, 2015). The main uses of WhatsApp among doctors are communication within clinical teams for discussion of medical cases, second opinion and tutoring, and communications between patients and doctors (Natarajan and Nair, 2015; Sarode and Sarode, 2016; Nardo et al., 2016; Johnston et al., 2015). There are works describing the use of WhatsApp in oral medicine (Petruzzi and De Benedittis, 2016), radiology (Giordano et al., 2015), emergency surgery (Nardo et al., 2016), orthopedics (Khanna et al., 2015), engagement to diabetes treatment (Blackstock et al., 2016), among others. The authors affirm that the communication facilitated by the tool can increase the productivity of the healthcare teams, enhance the tutoring of junior team members by senior staff, and reduce the times nurses spend trying to communicate with clinicians. An interesting possibility raised by one of the works is the use of communication records for audit and training purposes (Johnston et al., 2015). Yet the authors point out the concerns about the need to protect sensitive data, and the need to follow policies to approach confidentiality and safety of patient data.

These are sensitive points that comprise the great weakness of WhatsApp use in healthcare, although the company is planning to release second semester of 2016) a complete point-to-point encryption (WhatsApp, 2016). Some authors raised these points in more detail (Natarajan and Nair, 2015; Johnston et al., 2015). Some important problems mentioned are: (1) possibility of access to the data during transmission, even after the introduction of a partial encryption scheme by the development company; (2) lack of patient's consent for sharing their images over the internet in the cases of clinical teams discussion; (3) clear characterization of teleconsultation when patients sent their data to the doctors. Concerning this last point, all authors agree that "online consult" or "teleconsultation" is no substitute for a real clinical examination, since the patients cannot be clear about symptoms or images can have inappropriate resolution for diagnosis. Moreover, legal actions are applicable to errors that may result from these consults.

To cope with these problems, healthcare authorities of some countries have published recommendations regarding the use of social media that can be applied to the use of WhatsApp as well. The next section presents some of these recommendations. The severity on the application of these recommendations, or even the lack of a national recommendation, may well explain the big differences among the countries in the adoption of WhatsApp use, reported at the beginning of this section. Based on these recommendations, some alternative solutions are being delivered, such as lua (https://www.getlua.com/) and DocbookMD (https://www.docbookmd.com). Both are HIPAA compliant messaging applications with the same features of WhatsApp.

## Challenges

The collaborative nature of the communication between healthcare communities is one main advantage of using social media platforms. Yet, since healthcare is a sensitive issue, the adoption of these new channels of communication has to be performed with extreme care by all actors involved.

A primary concern when discussing healthcare use of social media is the reliability of information. Given the speed at which information spreads throughout the Internet, posts with unreliable or doubtful information may lead to misinformation and present a potential threat of inappropriate people's behavior. An example is given in a paper of Fernandez-Luque and Bau (2015) relating the recent outbreak of Ebola in West Africa. According to many studies, most of the messages contained false information about the spreading of the disease, raising the level of panic, and helped to spread fake treatments. Other disturbing examples are the communities that promote eating disordered behavior and antivaccine movements (Kahn, 2008; Fernandez-Luque and Bau, 2015).

Another major question relates to privacy and confidentiality of private healthcare data shared at the Internet. The question is particularly relevant for healthcare professionals interacting by social media. Main potential risks of healthcare professionals when using social media are (Ibarra-Yruegas et al., 2015): (1) loss of patient confidence in the professional relationship; (2) public exposure of patient private information, action that can be punished by law; (3) professionalism mistrust by publication of improper material or false information; (4) loss of professional-social boundaries.

To guide the conduct of healthcare professionals when using these media, there are published policies intended to help them to avoid the above-mentioned risks. The National Council of State Boards of Nursing published regulatory concerns and guidelines for nurses (Cronquist and Spector, 2011), The American College of Physicians and the Federation of State Medical Boards have published a position paper with recommendations about sensitive issues involved in the use of social media (Farnan et al., 2013). In the United States, compliance to the HIPAA strongly influences the social media actions of healthcare institutions and professionals.

Particularly concerning WhatsApp use, The National Health Service in England published a severe recommendation stating that "whatever the other merits of WhatsApp, it should never be used for the sending of information in the professional healthcare environment" (Notholt, 2015).

Still, patients are using the technology anyway. In spite of these problems, social media have indeed created news ways of communication between healthcare actors, delivering healthcare information and improving medical learning. Institutions, professional boards and governments around the world should not only create conduct guidelines, but also assure that the professionals follow them. In addition, Legislation has to follow this trend, in order to regulate something that is already happening.

## CONCLUSIONS

Healthcare is an information-based area with increasing use of ICT. In this scenario, EHR as a tool to collect and store patient data and to supply this information to healthcare professionals on request seems to be the first step to the integration of healthcare system services and has demonstrated a growing interest of care providers, patients, payers and policy makers around the World. However the adoption of EHR in the world differs in developing and developed countries.

Health information systems have to be developed and explored that enhances opportunities for global access to health services and medical knowledge. Ubiquitously available computing resources and networks existing worldwide for the transmission of all varieties of data, will allow us to consider new types of information systems for healthcare, including new kinds of health monitoring and also new opportunities for the analysis of biomedical and health data. These multiinstitutional information system architectures and infrastructures will, when appropriately designed and managed, provide new opportunities for the whole field of biomedical and health informatics as well as of biomedical statistics and epidemiology.

These new opportunities for the systematic processing of data, information, and knowledge in medicine and healthcare may considerably contribute to the progress of medicine and the health sciences as well as to the progress of informatics in general.

However the desired HIE has not been reached yet but there are important advances in technology and the level of maturity of hospitals and healthcare institutions is increasing, indicating a future integration of systems with interoperability, which may take longer than expected. The main thrust for the definitive adoption of HIE is to provide cost-effective sustainability, otherwise companies will not adopt willingly such initiatives.

Telemedicine presents many challenges and encompasses a great variety of activities. Certain activities are more prone to success than others are, and those are more likely to remain. On the other hand, there are important areas of telemedicine, such as remote assistance in case of calamity and teleeducation, which will require the active support of the governments. International nonprofit organizations for healthcare are essential to support developing countries with severe economic or social problems (such as war zones). The continuous development of new technologies is important to lower the cost of telemedicine (such as cheap smartphones and sensors) and increase the inclusion of people so they can have a better care. A special case of Telemedine involving mobile devices and its corresponding infrastructure, named mHealth can bring many benefits to private and public health. Yet, there are several challenges to overcome before the promises offered by the new technology can be reached. The next step in the development of mHealth projects and solutions is to face the main issues related to their widespread adoption. Important items that will help scaled deployment of these initiatives include healthcare data security, the adoption of healthcare standards, the integration to national and international health programs, the creation of a robust evidence base of the clinical benefits, and the conception of means for sustainable implementation and maintenance of the solution in large scale.

Social media have created news ways of communication between healthcare actors, delivering healthcare information and improving medical learning. There are, however, many risks involved in this communication media such as the reliability of information, privacy and confidentiality of the data, and misconduct of healthcare professionals. Institutions, professional boards, and governments around the world are creating guidelines to regulate the use of these media. In addition, Legislation has to follow this trend, in order to regulate something that is already happening.

# References

Agarwal, S., et al., 2015. Evidence on feasibility and effective use of mHealth strategies by front-line health workers in developing countries: systematic review. Tropic. Med. Int. Health 20 (8), 1003–1014.

Agha, L., 2014. The effects of health information technology on the costs and quality of medical care. J. Health Econ. 34, 19–30.

Akter, S., Ray, P., 2010. mHealth—an ultimate platform to serve the unserved. Yearbk. Med. Inform. 94–100.

Alam, M., et al., 2010. Assessing the scope for use of mobile based solution to improve maternal and child health in Bangladesh: a case study on efficiency of community health workers, automated risk assessment of patients and web based data collection. ICTD 2010 London Conference. December 2010, Royal Holloway, University of London.

Antheunis, M.L., Tates, K., Nieboer, T.E., 2013. Patients' and health professionals' use of social media in health care: motives, barriers and expectations. Patient Educ. Counsel. 92 (3), 426–431. Available at: <http://dx.doi.org/10.1016/j.pec.2013.06.020>.

Aranda-Jan, C.B., Mohutsiwa-Dibe, N., Loukanova, S., 2014. Systematic review on what works, what does not work and why of implementation of mobile health (mHealth) projects in Africa. BMC Public Health 14 (1), 188. Available at: <http://www.pubmedcentral.nih.gov/articlerender.fcgi?artid=3942265&tool=pmcentrez&rendertype=abstract/n. http://www.biomedcentral.com/1471-2458/14/188>.

Armstrong-Wells, J., et al., 2009. Prevalence and predictors of perinatal hemorrhagic stroke: results from the kaiser pediatric stroke study. Pediatrics 123 (3), 823–828.

ASTM & HL7, 2007. Continuity of Care Document.

Baig, M.M., GholamHosseini, H., Connolly, M.J., 2015. Mobile healthcare applications: system design review, critical issues and challenges. Australas. Phys. Eng. Sci. Med. 38 (1), 23–38. Available at:<http://link.springer.com/10.1007/s13246-014-0315-4>.

Ballesteros, S., Mayas, J., 2015. Selective attention affects conceptual object priming and recognition: a study with young and older adults. Front. Psychol. 6 (JAN), 2014–2016.

Barber, B., 1998. Patient data and security: an overview. Int. J. Med. Inform. 49 (1), 19–30.

Bates, D.W., Bitton, A., 2010. The future of health information technology in the patient-centered medical home. Health Aff. 29 (4), 614–621.

Bediang, G., et al., 2014. The RAFT telemedicine network: lessons learnt and perspectives from a decade of educational and clinical services in low- and middle-incomes countries. Front. Public Health 2 (October), 180.

Bellos, C., et al., 2011. CHRONIOUS: a wearable platform for monitoring and management of patients with chronic disease. 2011 Annual International Conference of the IEEE Engineering in Medicine and Biology Society. IEEE, pp. 864–867. Available at: <http://ieeexplore.ieee.org/lpdocs/epic03/wrapper.htm?arnumber=6090225>.

Bender, J.L., Jimenez-Marroquin, M.C., Jadad, A.R., 2011. Seeking support on facebook: a content analysis of breast cancer groups. J. Med. Internet Res. 13 (1).

Ben-Zeev, D., et al., 2015. Strategies for mHealth Research: lessons from 3 Mobile intervention studies. Admin. Policy Ment. Health Ment. Health Serv. Res. 42, 157–167.

Berkowitz, S.A., et al., 2014. Low socioeconomic status is associated with increased risk for hypoglycemia in diabetes patients: the Diabetes Study of Northern California (DISTANCE). J. Health Care Poor Underserved 25 (2), 478–490.

Blackstock, S., et al., 2016. G534 The use of a WhatsApp™ broadcast group to improve knowledge and engagement of adolescents with type 1 diabetes. Arch. Dis. Child. 101 (Suppl. 1), A315–A316. Available at: <http://adc.bmj.com/lookup/doi/10.1136/archdischild-2016-310863.521>.

Boulos, M.N.K., et al., 2014. Mobile medical and health apps: state of the art, concerns, regulatory control and certification. Online J. Public Health Inform. 5 (3), 229.

Boyd, D.M., Ellison, N.B., 2008. Social network sites: definition, history, and scholarship. J. Comput. Mediat. Commun. 13, 210–230.

Braa, J., Monteiro, E., Sahay, S., 2004. Networks of action: sustainable health information systems across developing countries 1. Sustain. Health Inform. Syst. MIS Quarterly 28 (3), 337–362.

Cafazzo, J.A., et al., 2012. Design of an mHealth app for the self-management of adolescent type 1 diabetes: a pilot study. J. Med. Internet Res. 14 (3), e70.

Calgary Scientific, 2013. ResolutionMD.

Campanella, P., et al., 2016. The impact of electronic health records on healthcare quality: a systematic review and meta-analysis. Eur. J. Public Health 26 (1), 60–64.

Cao, F., Huang, H.K., Zhou, X.Q., 2003. Medical image security in a HIPAA mandated PACS environment. Comput. Med. Imaging Graph. 27 (2-3), 185–196.

Carr, R., Wootton, R., 2013. Telemedicine for altruistic reasons: the Swinfen Charitable Trust. Health Executives 3–7.

Carroll, A.E., et al., 2011. Using a cell phone-based glucose monitoring system for adolescent diabetes management. Diabetes Educ. 37 (1), 59–66.

CEN, 1993. CEN/TC 251—Health Informatics.

CGI.br, 2014. Survey on the Use of Information and Communication Technologies in Brazilian Healthcare Facilities—ICT in Health.

Chaiyachati, K.H., et al., 2013. A pilot study of an mHealth application for healthcare workers: poor uptake despite high reported acceptability at a rural South African community-based MDR-TB treatment program. PLoS ONE 8 (5), 1–8.

Chen, C., et al., 2012. Making sense of mobile health data: an open architecture to improve individual- and population-level health. J. Med. Internet Res. 14 (4), 1–10.

Cheuk, B.L.Y., Cheung, G.C.Y., Cheng, S.W.K., 2004. Epidemiology of venous thromboembolism in a Chinese population. Br. J. Surg. 91 (4), 424–428.

Chiarini, G., et al., 2013. mHealth technologies for chronic diseases and elders: a systematic review. IEEE J. Sel. Areas Commun. 31 (9), 6–18.

Cho, I., et al., 2010. Design and implementation of a standards-based interoperable clinical decision support architecture in the context of the Korean EHR. Int. J. Med. Inform. 79 (9), 611–622.

Chretien, K.C., Kind, T., 2013. Social media and clinical care: ethical, professional, and social implications. Circulation 127 (13), 1413–1421.

Conroy, D.E., Yang, C.H., Maher, J.P., 2014. Behavior change techniques in top-ranked mobile apps for physical activity. Am. J. Prev. Med. 46 (6), 649–652. Available at: <http://dx.doi.org/10.1016/j.amepre.2014.01.010>.

Coppola, F., et al., 2016. Results of an Italian survey on teleradiology. Radiol. Med. 121 (8), 652–659.

Croen, L.A., Grether, J.K., Yoshida, C.K., Odouli, R., Van de Water, J., 2005. Maternal autoimmune diseases, asthma and allergies, and childhood autism spectrum disorders: a case-control study. Arch. Pediatr. Adolesc. Med. 159 (2), 151–157.

Cronquist, R., Spector, N., 2011. Nurses and social media: regulatory concerns and guidelines. J. Nurs. Regul. 2 (3), 37–40. Available at: <http://linkinghub.elsevier.com/retrieve/pii/S2155825615302659>.

Crul, S., 2014. The mHealth opportunity in Sub-Sahara Africa: the path towards practical application. Deloitte .

David, G., et al., 2013. Economic measurement of medical errors using a hospital claims database. Value Health 16 (2), 305—310.

de la Torre-Díez, I., et al., 2015. Cost-utility and cost-effectiveness studies of telemedicine, electronic, and mobile health systems in the literature: a systematic review. Telemed. e-Health 21 (2), 81—85.

Delpierre, C., et al., 2004. A systematic review of computer-based patient record systems and quality of care: more randomized clinical trials or a broader approach?. Int. J. Qual. Health Care 16 (5), 407—416.

Dhukaram, A.V., et al., 2011. End-user perception towards pervasive cardiac healthcare services: benefits, acceptance, adoption, risks, security, privacy and trust. In: 2011 5th International Conference on Pervasive Computing Technologies for Healthcare (PervasiveHealth) and Workshops, pp. 478—484.

Dixon, A.K., FitzGerald, R., 2008. Outsourcing and teleradiology: potential benefits, risks and solutions from a UK/European perspective. J. Am. Coll. Radiol. 5 (1), 12—18.

Drazen, E., Lorincz, C., Dahlweid, M., 2012. Should healthcare organizations use social media? CSR World Magazine Summer 2012.

Duarte, G. et al., 2010. Borboleta and SaguiSaúde—open source mobile telehealth for public home Healthcare. In: Proceedings of the 8th International eHealth, Telemedicine and Health ICT Forum (Med-e-Tel).

Duchesne, J.C., et al., 2008. Impact of telemedicine upon rural trauma care. J. Trauma 64 (1), 92—98.

Duke, J.R., Bowers, G.H., 2006. Scope and sites of electronic health record systems. In: Lehmann, H., et al., (Eds.), Aspects of Electronic Health Record Systems. Springer Science-Business Media Inc, New York, NY, p. 483.

EN 13606 Association, 2010. CEN/ISO 13606 Standard.

European Commission, 2011. Interoperable eHealth Is Worth it Securing Benefits From Electronic Health Records and ePrescribing.

European Commission, 2016. Code of conduct on privacy for mHealth apps. Available at: <https://ec.europa.eu/digital-single-market/en/news/code-conduct-privacy-mhealth-apps-has-been-finalised> (accessed 27.07.16).

Farnan, J.M., et al., 2013. Position paper annals of internal medicine online medical professionalism: patient and public relationships: policy statement from the American College of Physicians and the Federation of State Medical Boards. Ann. Intern. Med. 158 (October 2012), 620—627. Available at: <http://annals.org/article.aspx?articleid=1675927>.

Fernandez-Luque, L., Bau, T., 2015. Health and social media: perfect storm of information. Healthc. Inform. Res. 21 (2), 67—73.

Ferrara, F.M., Sottile, P.A., 1999. The holistic architectural approach to integrating the healthcare record in the overall information system. Stud. Health Technol. Inform. 68, 847—852.

Food and Drug Administration/US Department of Health and Human, 2015. Guidance for industry and food, p.7. Available at: <http://www.fda.gov/downloads/MedicalDevices/.../UCM263366.pdf>.

Forson, P.K., et al., 2013. The use of open source electronic medical records in an urban ED in Kumasi-Ghana. Afr. J. Emerg. Med. 3 (4, Suppl.), S14.

Franz-Vasdeki, J., et al., 2015. Taking mHealth solutions to scale: enabling environments and successful implementation. J. Mobile Technol. Med. 4 (1), 35—38. Available at: <http://www.journalmtm.com/2015/taking-mhealth-solutions-to-scale-enabling-environments-and-successful-implementation/>.

Frost, J.H., Massagli, M.P., 2008. Social uses of personal health information within PatientsLikeMe, an online patient community: what can happen when patients have access to one another's data. J. Med. Internet Res. 10 (3), e15, February 2008. <http://dx.doi.org/10.2196/jmir.1053>.

Furuie, S.S., et al., 2006. Managing medical images and clinical information: InCor's experience. IEEE Trans. Inform. Technol. Biomed. 11 (5), 17−24.

Galliber, J.M., et al., 2008. Paper forms with PDA forms in an office-based patient survey. Ann. Fam. Med. 6 (2), 154−160. Available at: http://www.pubmedcentral.nih.gov/articlerender.fcgi?artid = 2267414&tool = pmcentrez&rendertype = abstract

Ganapathy, K., Ravindra, A., 2008. mHealth: a potential tool for health care delivery in India. In Making the eHealth Connection, Bellagio, Italy, July 13−August 8, 2008, pp. 1−9. Available at: <http://www.ehealth-connection.org/files/conf-materials/mHealth_A%20potential%20tool%20in%20India_0.pdf>.

Ganeshan, M., Prashanta, S., Marry, V., 2011. The use of mobile phone as a tool for capturing patient data in southern rural Tamil Nandu India. J. Health Inform. Dev. Countries 5, 9.

Gibbons, P., et al., 2007. Coming to terms: scoping interoperability for healthcare. In: Health Level Seven EHR Interoperability Work Group, February 7, 2007.

Giordano, V., et al., 2015. WhatsApp messenger is useful and reproducible in the assessment of tibial plateau fractures: inter- and intra-observer agreement study. Int. J. Med. Inform. 84 (2), 141−148. Available at: <http://dx.doi.org/10.1016/j.ijmedinf.2014.11.002>.

Grajales, F.J., et al., 2014. Social media: a review and tutorial of applications in medicine and health care. J. Med. Internet Res. 16 (2).

Griffis, H.M., et al., 2014. Use of social media across US hospitals: descriptive analysis of adoption and utilization. J. Med. Internet Res. 16 (11), 1−11.

Van Der Haak, M., et al., 2003. Data security and protection in cross-institutional electronic patient records. Int. J. Med. Inform. 70 (2-3), 117−130.

Halton, J., et al., 2014. Teleradiology usage and user satisfaction with the telemedicine system operated by médecins sans frontières. Front. Public Health 2 (October), 202.

Hamm, M.P., et al., 2013. Social media use among patients and caregivers: a scoping review. BMJ Open 3 (5), 1−10. Available at: http://www.pubmedcentral.nih.gov/articlerender.fcgi?artid = 3651969&tool = pmcentrez&rendertype = abstract.

Hampshire, K., et al., 2015. Informal m-health: how are young people using mobile phones to bridge healthcare gaps in Sub-Saharan Africa? Soc. Sci. Med. 142, 90−99.

Hardcastle, T.C., Oteng, R., 2011. Trauma care in Africa: triumphs and challenges. Afr. J. Emerg. Med. 1 (2), 53−54.

Harris, P.A., et al., 2009. Research electronic data capture (REDCap)-A metadata-driven methodology and workflow process for providing translational research informatics support. J. Biomed. Inform. 42 (2), 377−381.

Health Level 7, 2015. HL7 Version 3 Product Suite.

Hesdorffer, D.C., et al., 2012. Epilepsy, suicidality, and psychiatric disorders: a bidirectional association. Ann. Neurol. 72 (2), 184−191.

Hess, D.C., et al., 2005. REACH: clinical feasibility of a rural telestroke network. Stroke 36 (9), 2018−2020.

Hibbard, J.U., et al., 2010. Respiratory morbidity in late preterm births. JAMA 304 (4), 419−425.

HIMSS, 2016a. Asia Pacific EMR Adoption Model. Healthcare Information and Management Systems Society.

HIMSS, 2016b. European Countries EMR Adoption Model. Healthcare Information and Management Systems Society.

HIMSS, 2016c. United States EMR Adoption Model. Healthcare Information and Management Systems Society.

HL7, 1989. Health Level 7.

HL7, 2007. Clinical Document Architecture.

Hoerbst, A., Ammenwerth, E., 2010. Electronic health records: a systematic review on quality requirements. Methods Inform. Med. 49 (4), 320−336.

Huckvale, K., et al., 2015. Smartphone apps for calculating insulin dose: a systematic assessment. BMC Med. 13 (1), 106. Available at: <http://www.biomedcentral.com/1741-7015/13/106>.

Iakovidis, I., 1998. Towards personal health record: current situation, obstacles and trends in implementation of electronic healthcare record in Europe. Int. J. Med. Inform. 52 (1-3), 105−115.

Ibarra-Yruegas, B.E., et al., 2015. Social networks in medical practice. Med. Universit. 17 (67), 108−113. Available at: <http://linkinghub.elsevier.com/retrieve/pii/S1665579615000332>.

IEEE, 1991. IEEE Standard Computer Dictionary: a compilation of IEEE Standard computer glossaries. IEEE Stand. 610, 1−217.

IHE, 1998. Integrating the Healthcare Enterprise.

IHE, 2006. Cross-Enterprise Document Sharing.

Institute of Medicine, 1991. In: Dick, R.S., Steen, E.B. (Eds.), The Computer-Based Patient Record: An Essential Technology For Health Care. National Academic Press, Washington, D.C.

Institute of Medicine, 2003. *Key Capabilities of an Electronic Health Record System*. Committee on Data Standards for Patient Safety. Washington, D.C.

International Health Terminology Standards Development Organisation, 2002. SNOMED CT.

ISO, 1998. ISO/TC 215 Health informatics.

ISO, CEN & IEEE, 2004. IEEE Standards—Health informatics.

ITU, 2015. The World in 2015: Facts and Figures. International Telecommunications Union.

Iwaya, L.H., et al., 2013. Mobile health in emerging countries: a survey of research initiatives in Brazil. Int. J. Med. Inform. 82 (5), 283−298. Available at: <http://dx.doi.org/10.1016/j.ijmedinf.2013.01.003>.

J&H MEDSOFT, 2016. DocbookMD.

Jahan, S., Mozammel Hoque Chowdhury, M., 2014. mHealth: a sustainable healthcare model for developing world. Am. J. Model. Optim. 2 (3), 73−76. Available at: <http://pubs.sciepub.com/ajmo/2/3/2/index.html>.

Javelot, H., et al., 2014. Telemonitoring with respect to mood disorders and information and communication technologies: overview and presentation of the PSYCHE project. BioMed Res. Int. 2014, 104658.

Jha, A.K., et al., 2008. The use of health information technology in seven nations. Int. J. Med. Inform. 77 (12), 848−854.

Jian, W.-S., et al., 2007. Building a portable data and information interoperability infrastructure—framework for a standard Taiwan Electronic Medical Record Template. Comput. Methods Programs Biomed. 88 (2), 102−111.

Johnston, M.J., et al., 2015. Smartphones let surgeons know WhatsApp: an analysis of communication in emergency surgical teams. Am. J. Surg. 209 (1), 45−51.

Kaewkungwal, J., et al., 2010. Application of smart phone in "Better Border Healthcare Program": a module for mother and child care. BMC Med. Inform. Decis. Mak. 10, 69. Available at: <http://www.pubmedcentral.nih.gov/articlerender.fcgi?artid = 2989931&tool = pmcentrez&rendertype = abstract>.

Kahn, J., 2008. The Wisdom of Patients: Health Care Meets Online Social Media . Available at: <http://search.ebscohost.com/login.aspx?direct = true&db = bth&AN = 32816528&site = ehost-live&scope = site>.

Kanaya, A.M., et al., 2011. Heterogeneity of diabetes outcomes among Asians and Pacific Islanders in the US. Diabetes Care 34 (4), 930−937.

Kangarloo, H., et al., 2000. Improving the quality of care through routine teleradiology consultation. Acad. Radiol. 7 (3), 149−155.

Katz, D., et al., 2016. Questioning the reflection paradigm for diabetes mobile apps conference item. In: EAI International Conference on Wearables in Healthcare, 14−15 June 2016, Budapest, European Alliance for Innovation.

Khanna, V., et al., 2015. WhatsApp"ening in orthopedic care: a concise report from a 300-bedded tertiary care teaching center. Eur. J. Orthop. Surg. Traumatol. 25 (5), 821−826. Available at: <http://dx.doi.org/10.1007/s00590-015-1600-y>.

Kohane, I.S., et al., 1996. Sharing electronic medical records across multiple heterogeneous and competing institutions. Proc. AMIA Annu. Fall Symp. 608−612.

Köhler, F., et al., 2005. Partnership for the heart: German-Estonian health project for the treatment of congenital heart defects in Estonia. Health Policy 73 (2), 151−159.

Kolbe, N., et al., 2015. Point of care ultrasound (POCUS) telemedicine project in rural Nicaragua and its impact on patient management. J. Ultrasound 18 (2), 179−185.

König, H., 2005. Access to persistent health information objects: exchange of image and document data by the use of DICOM and HL7 standards. Int. Congress Series 1281, 932−937.

Kumar, G., et al., 2013a. The costs of critical care telemedicine programs. Chest 143 (1), 19−29.

Kumar, S., et al., 2013b. Mobile health technology evaluation: the mHealth evidence workshop. Am. J. Prev. Med. 45 (2), 228−236. Available at:<http://dx.doi.org/10.1016/j.amepre.2013.03.017>.

Kumar, N., et al., 2015. A content analysis of smartphone-based applications for hypertension management. J. Am. Soc. Hypertens. 9 (2), 130−136. Available at: <http://dx.doi.org/10.1016/j.jash.2014.12.001>.

Labrique, A., 2013. 12 Common applications and a visual framework. Glob. Health Sci. Pract. 1 (2), 1−12.

Laraia, B.A., et al., 2012. Place matters: neighborhood deprivation and cardiometabolic risk factors in the Diabetes Study of Northern California (DISTANCE). Soc. Sci. Med. 74 (7), 1082−1090.

Leon, N., Schneider, H., Daviaud, E., 2012. Applying a framework for assessing the health system challenges to scaling up mHealth in South Africa. BMC Med. Inform. Decis. Mak. 12, 123. Available at: <http://www.pubmedcentral.nih.gov/articlerender.fcgi?artid = 3534437&tool = pmcentrez&rendertype = abstract>

Lua Technologies, 2016. LUA.

Macedo Pinho Rotzsch, J., et al., 2005. Brazilian national standard for electronic form interchange among private health plan and healthcare providers. Stud. Health Technol. Inform. 977−982.

Macleod, B., et al., 2012. The architecture of a software system for supporting community-based primary health care with mobile technology: the Mobile Technology for Community Health (MoTeCH) initiative in Ghana. Online J. Public Health Inform. 4 (1), 1−17. Available at: <http://www.pubmedcentral.nih.gov/articlerender.fcgi?artid = 3615808&tool = pmcentrez&rendertype = abstract>

Mahajan, R., et al., 2014. Mortality among persons in care with hepatitis C virus infection: the Chronic Hepatitis Cohort Study (CHeCS), 2006−2010. Clin. Infect. Dis. 58 (8), 1055−1061.

Männistö, T., et al., 2015. Acute air pollution exposure and blood pressure at delivery among women with and without hypertension. Am. J. Hypertens. 28 (1), 58–72.

Mannu, P., 2015. The Digital Health Debate: A Report on How Doctors Engage with Digital technology in The Workplace. Cello Health Insight.

Manousos, D., et al., 2013. First results about the use of a patient portal by people with diabetes in a rural area. 2013 E-Health and Bioengineering Conference (EHB). IEEE, pp. 1–5. Available at: <http://ieeexplore.ieee.org/lpdocs/epic03/wrapper.htm?arnumber = 6707375>.

Marcolino, M.S., et al., 2013. Implementation of the myocardial infarction system of care in city of Belo Horizonte, Brazil. Arq. Bras. Cardiol. 100 (4), 307–314.

Marcus, D.S., et al., 2007. The extensible neuroimaging archive toolkit. Neuroinformatics 5 (1), 11–33.

Mariño, R., et al., 2016. Cost-analysis of teledentistry in residential aged care facilities. J. Telemed. Telecare. 22 (6), 326–332.

Martínez-Pérez, B., De La Torre-Díez, I., López-Coronado, M., 2013. Mobile health applications for the most prevalent conditions by the World Health Organization: review and analysis. J. Med. Internet Res. 15 (6), .

Masi, C., et al., 2012. Using an established telehealth model to train urban primary care providers on hypertension management. J. Clin. Hypertens. 14 (1), 45–50.

Mechael, P., et al., 2010. Barriers and gaps affecting mHealth in low and middle income countries: Policy white paper. Available at: <http://crcrth670-f11-sullivan.wikispaces.umb.edu/file/view/mHealth_EARTH + Institute_Low + Mid + Income + Co's_Barriers_WhitePaper_Columbia + U + '10.pdf>.

Mechanic, D., 2008. Rethinking medical professionalism: the role of information technology and practice innovations. Milbank Quart. 86 (2), 327–358.

Microsoft, 2016. HealthVault.

Millard, P.S., Bru, J., Berger, C.A., 2012. Open-source point-of-care electronic medical records for use in resource-limited settings: systematic review and questionnaire surveys. BMJ Open 2 (4).

Kashif Mirza, M., Sajjad, Z., Yousuf, M., Usman, Z., 2008. Teleradiology between Afghanistan and Pakistan: one year experience. Pak. J. Radiol. 18 (1), 22–25.

Mobasheri, M.H., et al., 2014. Smartphone breast applications—what's the evidence? Breast 23 (5), 683–689. Available at: <http://dx.doi.org/10.1016/j.breast.2014.07.006>.

Mobile Communications for Medical Care, Final Report, April 21, 2011, University of Cambridge and China Mobile.

Moorhead, S.A., et al., 2013. A new dimension of health care: systematic review of the uses, benefits, and limitations of social media for health communication. J. Med. Internet Res 15 (4), 1–16.

Mookherji, S., et al., 2015. Unmet need: improving mhealth evaluation rigor to build the evidence base. J. Health Commun. 0730 (September 2015), 1–6. Available at:<http://www.tandfonline.com/doi/full/10.1080/10810730.2015.1018624>.

Moorman, A.C., et al., 2013. Baseline characteristics and mortality among people in care for chronic viral hepatitis: the Chronic Hepatitis Cohort Study. Clin. Infect. Dis. 56 (1), 40–50.

Mori, A.R., Consorti, F., 1998. Exploiting the terminological approach from CEN/TC251 and GALEN to support semantic interoperability of healthcare record systems. Int. J. Med. Inform. 48 (1-3), 111–124.

Muehlberg, F., et al., 2015. A multicenter cardiovascular MR network for tele-training and beyond: setup and initial experiences. J. Am. Coll. Radiol. 12 (8), 876–883.

Murphy, S.N., et al., 2006. Integration of clinical and genetic data in the i2b2 architecture. AMIA Annu. Symp. Proc. 2, 1040.

Nardo, B., et al., 2016. Optimizing patient surgical management using WhatsApp application in the Italian healthcare system. Telemed. e-Health , p.tmj.2015.0219. Available at: <http://online.liebertpub.com/doi/10.1089/tmj.2015.0219>.

Natarajan, S., Nair, A., 2015. Outsmarted by the smartphone!. Indian J. Ophthalmol. 63 (10), 757. Available at: <http://www.ijo.in/text.asp?2015/63/10/757/171502>.

NEMA, 1998. DICOM Homepage.

Notholt, S.A., 2015. 21 Information Governance Bulletin.

Nsanzimana, S., et al., 2012. Cell phone-based and internet-based monitoring and evaluation of the national antiretroviral treatment program during rapid scale-up in Rwanda. J. Acquir. Immune Defic. Syndr. 59 (2), e17−e23.

OMG, 1997. CORBAmed roadmap.

openEHR, 2006. Archetype Technology Overview.

Osborn, D.P.J., et al., 2015. Cardiovascular risk prediction models for people with severe mental illness: results from the prediction and management of cardiovascular risk in people with severe mental illnesses (primrose) research program. JAMA Psychiatry 72 (2), 143−151.

Pagni, F., et al., 2011. Virtual surgical pathology in underdeveloped countries: the Zambia Project. Arch. Pathol. Lab. Med. 135 (2), 215−219.

Paré, G., et al., 2007. Systematic review of home telemonitoring for chronic diseases: the evidence base. J. Am. Med. Inform. Assoc. 14 (3), 269−277.

PatientView, 2016. *The Myhealthapps Directory 2015-2016*.

Petersen, C., DeMuro, P., 2015. Legal and regulatory considerations associated with use of patient-generated health data from social media and mobile health (mHealth) devices. Appl. Clin. Inform. 6 (1), 16−26. Available at: <http://www.pubmedcentral.nih.gov/articlerender.fcgi?artid = 4377557&tool = pmcentrez&rendertype = abstract>.

Petruzzi, M., De Benedittis, M., 2016. WhatsApp: a telemedicine platform for facilitating remote oral medicine consultation and improving clinical examinations. Oral Surg. Oral Med. Oral Pathol. Oral Radiol. 121 (3), 248−254. Available at:< http://dx.doi.org/10.1016/j.oooo.2015.11.005>.

Pham, J.C., et al., 2012. Reducing medical errors and adverse events. Annu. Rev. Med. 63 (1), 447−463.

Pixmeo SARL, 2004. OsiriX Imaging Software.

Prgomet, M., Georgiou, A., Westbrook, J.I., 2009. The impact of mobile handheld technology on hospital physicians' work practices and patient care: a systematic review. J. Am. Med. Inform. Assoc. 16 (6), 792−801.

PWC & GSMA Connected Living Programme, 2013. Socio-Economic Impact of mHealth: An Assessment Report for the European Union.

Qin, R., et al., 2013. Reliability of a telemedicine system designed for rural Kenya. J. Prim. Care Commun. Health 4 (3), 177−181.

Qizilbash, N., et al., 2016. BMI and risk of dementia in two million people over two decades: a retrospective cohort study. Lancet Diabetes Endocrinol. 3 (6), 431−436.

Rassinoux, A.M., et al., 2002. XML as standard for communicating in a document-based electronic patient record: a three years experiment. Stud. Health Technol. Inform. 90, 673−678.

Regenstrief Institute, 1994. LOINC.

Ryu, S., 2012. Telemedicine: opportunities and developments in Member States: report on the Second Global Survey on eHealth 2009 (Global Observatory for eHealth Series, Volume 2). Healthc. Inform. Res. 18 (2), 153.

Sa, J.H.G., et al., 2016. Georeferenced and secure mobile health system for large scale data collection in primary care. Int. J. Med. Inform. 94, 91–99. Available at: <http://dx.doi.org/10.1016/j.ijmedinf.2016.06.013>.

Saliba, V., et al., 2012. Telemedicine across borders: a systematic review of factors that hinder or support implementation. Int. J. Med. Inform. 81 (12), 793–809.

Sarode, S., Sarode, G., 2016. WhatsApp use in dentistry: future prospects. J. Dental Res. Rev. 3 (1), 3. Available at: <http://www.jdrr.org/text.asp?2016/3/1/3/180104>.

Scott, R.E., Mars, M., 2015. Telehealth in the developing world: current status and future prospects. Smart Homecare Technol. TeleHealth 3, 25–37.

Selby, J., 1997. Linking automated databases for research in managed care settingse. Ann. Intern. Med. 127, 719–724.

Simonis, F., et al., 2011. NEPHRON + WEARABLE ARTIFICIAL KIDNEY. In: Abstracts: XXXVIII Annual ESAO & IV Biennial IFAO Congress, 9–12 October 2011, Porto, Portugal, p. 610.

Smith, E., Eloff, J.H., 1999. Security in health-care information systems—current trends. Int. J. Med. Inform. 54 (1), 39–54.

Sutton, L.N., 2011. PACS and diagnostic imaging service delivery—a UK perspective. Eur. J. Radiol. 78 (2), 243–249.

Thomas, S.L., et al., 2014. Chickenpox and risk of stroke: a self-controlled case series analysis. Clin. Infect. Dis. 58 (1), 61–68.

Thota, R.S., Divatia, J.V., 2015. WhatsApp: What an App!. Indian J. Crit. Care Med. 19 (6), 362–363. Available at: doi:10.4103/0972-5229.158284. <http://login.ezproxy.lib.vt.edu/login?url = http://search.ebscohost.com/login.aspx?direct=true&db=a9h&AN=103376000&scope=site>.

Tomlinson, M., et al., 2013. Scaling up mHealth: where is the evidence? PLoS Med. 10 (2), 1–5.

US Department of Labor, 2015. The Health Insurance Portability and Accountability Act.

United Nations Statistics Division, 2013. Standard Country and Area Codes Classifications. United Nations.

US National Library of Medicine, 1999. Medical Subject Headings.

US National Library of Medicine, 2009. Unified Medical Language System (UMLS).

Uslu, A.M., Stausberg, J., 2008. Value of the electronic patient record: an analysis of the literature. J. Biomed. Inform. 41 (4), 675–682.

Varshney, U., 2014. Mobile health: Four emerging themes of research. Decis. Support Syst. 66, 20–35. Available at: <http://dx.doi.org/10.1016/j.dss.2014.06.001>.

Versleijen, M., et al., 2015. A telegeriatric service in a small rural hospital: a case study and cost analysis. J. Telemed. Telecare 21 (8), 459–468.

Vigen, R., et al., 2013. Association of testosterone therapy with mortality, myocardial infarction, and stroke in men with low testosterone levels. JAMA 310 (17), 1829–1836.

WhatsApp, 2016. WhatsApp security. Available at: <www.whatsapp.com/security> (accessed 29.07.16).

Ward, M.M., Jaana, M., Natafgi, N., 2014. Systematic review of telemedicine applications in emergency rooms. Int. J. Med. Inform. 84 (9), 601–616.

Warmerdam, L., et al., 2012. Innovative ICT solutions to improve treatment outcomes for depression: the ICT4Depression project. Stud. Health Technol. Inform. 181 (September 2012), 339–343.

Webster, P.C., 2011. The rise of open-source electronic health records. Lancet 377 (9778), 1641–1642.

Weinstein, R.S., et al., 2014. Telemedicine, telehealth, and mobile health applications that work: opportunities and barriers. Am. J. Med. 127 (3), 183–187.

WHO, 1992. Classification of Diseases.

WHO, 2006. Electronic Health Records: Manual for Developing Countries. *World Health Organization*.

WHO Group Consultation on Health Telematics, 1998. A health telematics policy in support of WHO's Health-for-all strategy for global health development: report of the WHO Group Consultation on Health Telematics, 11–16 December, Geneva, 1997.

Widmer, R.J., et al., 2014. The augmentation of usual cardiac rehabilitation with an online and smartphone-based program improves cardiovascular risk factors and reduces rehospitalizations. J. Am. Coll. Cardiol. 63 (12), A1296. Available at: <http://linkinghub.elsevier.com/retrieve/pii/S0735109714612961>

Wong, W.S., et al., 2005. Outsourced teleradiology imaging services: an analysis of discordant interpretation in 124,870 cases. J. Am. Coll. Radiol. 2 (6), 478–484.

Wootton, R., 2008. Telemedicine support for the developing world. J. Telemed. Telecare 14 (3), 109–114.

World Health Organization, 2011. mHealth: new horizons for health through mobile technologies—Based on the findings of the second global survey on eHealth, Available at: <http://www.who.int/goe/publications/goe_mhealth_web.pdf>.

World Health Organization, 2012, The top 10 causes of death: the 10 leading causes of death by country income group (2012), Available at: < http://www.who.int/mediacentre/factsheets/fs310/en/index1.html >

Zaidan, A.A., et al., 2015. Challenges, alternatives, and paths to sustainability: better public health promotion using social networking pages as key tools. J. Med. Syst. 39 (2).

# CHAPTER 4

# Standards for Global Health Information Systems

**W. Ed Hammond**

Duke University School of Medicine, Durham, NC, United States

## CONTENTS

Standards are the underlying enabler for communications and interoperability. The development of standards has been primarily driven by the requirements of the market place and have evolved over time. The first appearance of standards to support the development of systems in healthcare began in the 1970s. I will use the term health data standards to refer to standards created broadly for this purpose. In the academic setting, papers about health data standards began to appear in the early 1980s. Some of the first standards addressed the requirements of clinical laboratory systems. The creation of formal organizations, called Standard Developing Organizations (SDOs), began in the late 1980s. A number of SDOs appeared over a short period of time. Most of these SDOs had a specific purpose for being and did not initially appear to be competitive. The necessity of addressing more requirements brought these SDOs into a competitive mode. Then as well as now, a favorite statement is "The thing I like best about standards is that there are so many to choose from." Many existing standards serve the same purpose but do it differently. Users are stuck with selecting one standard and mapping to different standards that are used by others. Not only is the mapping always imperfect, but there is a continuing cost, and synchronization among standards is impossible. A challenge for the future is to select a single standard for each purpose and retire the others.

## THE KEY SDO ORGANIZATIONS

This section identifies the standards bodies that are engaged in some aspects of global standardization. Other standards bodies exist but are not part of the global space. A more detailed description of SDOs may be found in Shortliffe and Cimino (2012). The first organizations discussed originated in the United States followed by organizations formed in other countries.

Health data standards from a clinical research perspective are discussed Richesson and Andrews (2006).

The American College of Radiology (ACR) and the National Electrical Manufacturers Association (NEMA) created one of the first imaging standards known as DICOM (dicom.nema.org, 2016) (Digital Imaging and Communications in Medicine), published in 1985. DICOM later became the name of the SDO itself as it morphed into an international organization, and ACR/NEMA became a member of that organization. DICOM today is used throughout the world and is the dominant if not only standard addressing imaging.

ASTM International (www.astm.org/COMMITTEE/E31.htm, 2016), founded in 1898, is one of the oldest SDOs in the world. ASTM creates standards in many areas, and in 1970 formed the Technical Committee E31 to address standards related to security and communications of data. The first health data standards was published in 1984 for transferring clinical observations between independent systems. ASTM created the Continuity of Care Record (CCR) in the early 2000s that has and is influencing the definition of a summary patient record for referrals or the transfer of a patient into another setting. Although CCR has limited use in itself, it became the input to a more formal data standard developed by Health Level Seven (www.hl7.org, 2016) (HL7) called the Continuity of Care Document (CCD). The CCD was based on another HL7 standard known as the Clinical Document Architecture (CDA).

HL7 was formed in 1987, and its initial focus was in the creation of standards to support a "Best of Breed" Hospital Information System (HIS). These early HISs were created from disparate component parts that were connected together to form an integrated system. Initially an interface had to be created between each of the components resulting in an excessive cost for this approach. A common standard should drive these interfacing costs downward, and they did. The competition were the larger vendors who offered a single system that addressed all needs. As a result the take-up by the industry was very slow. This HL7 standard, known as version 2.n, continues in use today. It is widely used throughout the world for the transporting of data. Other HL7 standards will be discussed later in this chapter, since they have an impact on global standards. HL7 became international with affiliate agreements with Germany and The Netherlands in the early 1990s.

The Institute of Electrical and Electronics Engineers (www.ieee.org, 2016) (IEEE) has created a family of standards that address the requirements of the Medical Information Bus (MIB). The MIB is designed to support interfaces for bedside devices and other similar devices into higher level systems. IEEE activities are currently harmonized with HL7 and The European Committee for Standardization (CEN).

The Integrating the Healthcare Enterprise (www.ihe.net, 2016) (IHE) was created in 1998 as a joint effort of the Radiological Society of North America and the Healthcare Information and Management Systems Society. Its primary reason for forming was not to create standards but to create a framework for identifying the sets of established standards necessary for end-to-end exchange of data. This framework, known as an implementation profile, defines the transactions needed to communicate images and patient data across multivendor environments. IHE has also become international with affiliates around the world.

The Clinical Data Interchange Standards Consortium (www.cdisc.org, 2016) (CDISC) was formed in 1997 to create data standards that enable system interoperability to improve medical research. CDISC, along with HL7, has created a data model to unify the data components supporting clinical research. CDISC is an international organization and has created key standards for clinical research.

CEN (www.cen.eu, 2016) was established in 1991. Technical Committee 251 (TC 251) was formed for the development of health data standards. The goal of TC 215 was to develop the group of standards to support communication among disparate health care systems. A significant standard that has influenced the field is Health Informatics—Electronic Health Communication (EN 13606). This standard defines an information architecture for communicating all or part of an Electronic Health Record (EHR) of a patient interoperably among disparate healthcare systems. CEN shares its standards through the International Standards Organization (ISO) as part of the Vienna Agreement.

In 1989, interested parties in Europe and the United States lead to the creation, within ISO, of Technical Committee 215 (TC 215)—Health Informatics (http://www.iso.org/iso/iso_technical_committee?commid = 54960, 2016). The United States represented by the American National Standards Institute was assigned the role of Secretariat. That role is currently filled by the American Health Information Management Association. ISO is an important organization in global standards because of its outreach into the global community. Furthermore, many countries require the use of ISO standards, if they exist.

OpenEHR Foundation (www.openEHR.org, 2016) is an international, online community formed to create EHRs based primarily on the CEN 13606 standard. The creators of the Foundation had considerable experience in Europe with an object-oriented approach to the EHR, and joined forces in Australia to create the organization. The group became formal around the turn of the century and has grown in membership, visibility, and products. OpenEHR is built upon a set of archtypes conforming to the 13606 standard.

The Joint Initiative Council (http://www.jointinitiativecouncil.org, 2016) (JIC) was formed in 2009 with the purpose of creating a single standard with collaborative participation by interested SDOs. The original members of JIC were ISO TC 215, CEN, and HL7 International. Membership has grown to include CDISC, DICOM, IHE International, IHTSDO, and GS1. There have been some successes using this model. A current project of this group is the Patient Summary Standards Set, which will be further discussed in this chapter.

GS1(www.gs1.org, 2016), for more than 35 years, has created a global supply chain data standards in multiple industries. In healthcare, these standards are designed to improve efficiency and quality and to reduce medical errors. Specific standards of importance to global health include a global location number, a global trade item number, and a global data synchronization network. The GS1 barcodes are widely used throughout healthcare.

## DATA NAMING AND REPRESENTATION

One of the biggest challenges in the exchange and use of data in any setting is the data representation. What do we call the data element, what are its attributes, and what does it mean to the source person of the data? The basic component of a data element is the name because this is what we use when we refer to the data. We have a choice of standards that we use to define the coding system we use as a controlled standard. The naming scheme is further complicated in that we identify the different coding systems as controlled vocabularies, terminologies, classifications, nomenclatures, and ontologies. Although we use these words interchangeably, their purpose may be slightly different. The problem is further complicated in that what we frequently store in an EHR is a local coding system or set of words. The text representing the data may be well-defined and structured, or it may be unstructured narrative, often referred to as free text. Each clinical specialty has its own preferred terminologies that may or not be derived from a common set.

There are a number of common datasets, but the problem is they are uncommon across global boundaries and across different groups. It is estimated that there are several hundred different terminology sets in use around the world. Some of the more common controlled terminology sets are described in the following section.

### Controlled Terminologies Organizations

The International Classification of Disease (ICD) (www.who.int/whosis/icd10/, 2016) is one of the best-known terminologies and is published by the World Health Organization. The purpose of this coding system is for reporting mortality and morbidity statistics. The current version, ICD-10, is in use throughout much of the world. The next version (ICD-11) is already in Beta test. ICD-10 consists of a core classification of a three-digit code followed by a decimal, then other digits to provide additional detail. This numbering scheme is mostly hierarchical in nature and may be coded to significant detail. ICD-10 is used frequently to code items in a patient's Problem List. Some countries expand the code set with a clinical modification to permit more specific disease and symptom coding. In the United States, ICD-10 codes are used for claims and billing.

International Health Terminology Standards Development Organization (IHTSDO) (www.ihtsdo.org, 2016) is an international organization owned and governed by 28 international member countries. IHTSDO owns and manages SNOMED CT and is a leading clinical terminology in the global space. Member countries have free distribution rights in their country. SNOMED CT is perhaps one of the most comprehensive terminology sets in the world.

International Classification of Primary Care (ICPC) (https://en.wikipedia.org/wiki/ICPC-2_PLUS, 2016) is used throughout the world for the documentation of encounters in primary care, except infrequent use in the United States. The coding system is published by the World Organization of National Colleges, Academies and Academic Associations of General Practitioners/Family Physicians (WONCA) along with the World Health Organization. ICPC has been translated into several languages.

Logical Observations, Identifiers, Names, and Codes (LOINC) (loinc.org, 2016) is an international standard for identifying clinical laboratory observations. LOINC is maintained by the Regenstrief Institute and is publically available at no charge. LOINC has two main parts: laboratory LOINC mainly for laboratory test data and clinical LOINC for clinical observations. Clinical LOINC contains a subdomain of document ontology, which codes types of clinical reports and documents. LOINC is used in more than 172 countries.

## Identification of Medicinal Products

One of the most challenging problems in semantic interoperability is the naming of medicinal products. Not only does each country have its own set of names and characteristics, but cross-country, the same drugs have different names and different drugs may have the same name. The problem is further complicated in that the desired attributes of drugs are often included in different datasets. In the United States, RxNorm (https://www.nlm.nih.gov/research/umls/rxnorm/, 2016), produced by the NLM, is a standardized nomenclature for clinical drugs that combines its ingredients, strengths, and form. In the United States RxNorm is becoming the drug code of choice. The National Drug Codes (NDC) produced by the Food and Drug Administration (FDA) is applied to all drug packages. The Veterans Health Administration's National Drug File-Reference Terminology (NDF-RT) (http://www.va.gov/TRM/StandardPage.asp?tid = 5221^,    2016)    includes adoption of mechanisms of action, physiologic effects, and chemical ingredient by structure hierarchy subsets used to describe pharmacologic class. The Daily Med was derived from the HL7 Structural Product Labeling (SPL) standard and supplies medication information in computer readable form.

The WHO Drug Dictionary is an international classification of drugs that provides proprietary drug names used in different countries. Included are all active ingredients, clinical substances, and Chemical Abstract numbers. Drugs are classified according to the Anatomical-Therapeutic-Chemical (ATC) classification, with cross-references to manufacturers and reference sources.

## THE MAKING OF STANDARDS

A reasonable question to ask is how we decide what standards we create. In most cases it is a decision from the technical stakeholders based on the assumptions about what the users will need. For HL7's version 2 series, the motivation and scope were defined by the need to create a HIS made of component parts supplied by different vendors. In this case the technical stakeholders understood what flow of data and what data content were required. The standard was designed for the inpatient setting. But quickly, the requirements spread to other settings (ambulatory care) and other topic domains. To the extent that these requirements could be met within the existing framework, additional fields were added to the appropriate segments. But that approach was not sufficient. A user defined segment—the Z segment—was defined, and the generality of the standard was defeated. In addition developers simply added whatever data elements they needed to exchange at the end of existing defined strings. The result was the loss of interoperability in general, but still v 2.n was a usable standard within a known community. An implementation guide could define specifically what was required, and constructed interoperability was reestablished within that community.

Within different SDOs, similar patterns were used to define the requirements for standards. Each community had specific reasons for creating standards, addressing a specific domain with a specific scope. But in all cases, each experienced scope creep. The exception to this rule is the part of DICOM that created the standard for the images. That standard has been stable over several decades. However the accompanying standards have had the same problems of an expanding scope. The problem then as now is how to accommodate the expanding scope. If a standard does not permit me to send the data elements I wish to send, I will either modify the standard (and making it a non-standard) or look for another solution.

The content of most health data standards have been driven by groups that had a specific need. Standards are made by a combination of stakeholders early on consisting mostly of vendors and healthcare providers—the builders of systems and the users. Success of a standard was based on how well each group understood the other, but the industry had the edge since they were the ones who built the systems. Within the industry, competition existed, in which the closer the standard supports what a specific vendor had implemented, the less expensive was conformance. One reason for a vendor engagement in the SDO was in self-defense of their existing system.

As the obvious standards were created and implemented, gaps in existing standards were realized. One global example of this dilemma were regulatory standards. Since most standards are created by a consensus process, and the work is done for volunteers, the challenge is to match the need to the interests of the volunteers. Although creating gap standards is important for overall interoperability, it has been difficult to recruit workers for these standards. This problem has become even more acute as what standards must address has continued to expand.

Today, most standards begin with a use case that defines the purpose of the standard. If possible, more than one related use case is useful. Multiple use cases provide an understanding of a total scenario, and a more generic standard is likely to result. Use cases by themselves, however, are inadequate to define the requirements and specifications of the standard. Data flows, information flows, and work flows need to be understood. The tool for accomplishing this is to create Domain Analysis Model (DAM). A DAM is an abstract representation of a use case, complete enough to allow visualization of work and data flow, data elements required, decision-making, and other options. DAMs are usually represented using Unified Modeling Language. One of the challenges of standard-making is to have an effective interface between the technical perspective of the required standard and the Subject Matter Experts that can describe from the user perspective what the standard can do. A complete DAM requires both static (class and instant diagrams) and dynamic (activity diagrams) semantics.

In the United States, the FDA introduced regulatory standards with HL7. These standards were Individual Case Safety Reports (ICSR), Structured Product Labeling (SPL), and annotated electrocardiogram (aECG). Globally, these standards were of interest to the European Medicines Agency (EMA) and the International Council for Harmonization (ICH). These development of one of these standards (ICSR) was moved into the JIC, and with the engagement of ICH, EMA, and FDA, a single standard was created.

## Global Standards

Most of the SDOs have become international. From a standards perspective, there may be a distinct difference in international standards and global standards. In this context international implies a sharing standards among countries for use within that country. The perception is that differences exist within each country—driven by culture, government regulations, process, policy, and other factors—and therefore standards must be modified to fit each country. Those standards, though derived from the same core standard, become realm-specific. This process introduces problems of maintenance and synchronization as these standards evolve. The reality in most cases is that many vendor products are using in multiple countries and prefer to not have to make country-specific changes. These perceived differences need to be closely examined to insure that the cost of that option is absolutely necessary and cannot be addressed with synonyms and process modifications. Global standards are meant to be used for cross-country purposes and have the same meaning across the borders.

Global standards are often defined by use cases that may be too contrived to be realistic. Global standards mean the leadership of the countries must be convinced in efficacy of just standards. Success means that the government must involve typically the minister of health, and the health system, which become the senders and receivers of health data.

# INTEROPERABILITY

Interoperability has become one of the most used words in health information technology. It has become increasingly important because of the many current initiatives that require the transport and sharing of data across multiple sites. There are many levels of interoperability, depending on the use case (Richesson and Andrews, 2006). There are also many factors and many stakeholders that must be aligned for interoperability to work. Security and privacy must be addressed in the sharing of data. Purpose and need must be defined. Value must justify cost. The initial definition of interoperability came from the IEEE and limited the definition to technical or functional interoperability and semantic interoperability. Although these remain important and still a challenge, other factors such as what data to send and when, identification of patients cross country borders, quality and province of data, and many other factors must be addressed.

One of the biggest issues in the exchange of data is the data representation. The many controlled vocabulary sets and the different purposes of each set is part of the problem. The lack of full attribute specification is another problem. The problem is further complicated by the localization of terminologies and in the differences in the permissible value sets of the data elements.

Even of common terminologies are used, there are many different ways of expressing the same concepts. In clinical research phenotype definitions for cohort identification are dominated by the many terminology codes used to identify the same diseases.

The distinction between controlled terminology sets and data elements is becoming blurred. Data elements include a fuller set of attributes and accommodate structure from a finely grained single term to more complex structures. Data elements permit an architecture that accommodates all the components of an item, whether present or not. For example, a data element representing blood pressure would include a format for systolic and diastolic values as well as body position, area in which pressure is measured, units, cuff size, and circumstances such as exercise under which measurement is taken. Data elements may also be computed such as a body mass index computed from a height and weight measurement.

Most data element sets are based on ISO/IEC 11179 Metadata Registry (MDR) standard (https://en.wikipedia.org/wiki/ISO/IEC_11179, 2016). Unfortunately, even though these common data models are based on a standard, the population of the MDR differs. Simple items such as gender, race, and ethnicity have different value sets. Fig. 4.1 below illustrates a rich set of attributes for a data element. Including purpose should differentiate between similar terms, even if their names are similar.

The Clinical Information Modeling Initiative, now a HL7 Work Group, is creating a shared repository of detailed clinical information models that will support multiple contexts in health data. Within the United States, The National Library of Medicine has an initiative called the Value Set Authority Center, which has created a registry that contains the vocabulary value sets contained in the 2014 Meaningful Use Clinical Quality Measures and a data element catalog that identifies data element names required for capture in EHRs certified under Office of the National Coordinator (ONC) standards. Patient-Centered Clinical Research Network (PCORNet) has created a common data model to be used to link and query across multiple sites for pragmatic clinical trials.

## The Role of Standards in Interoperability

Most of the new initiatives in the health and healthcare require the sharing and hence transport of data. The initiatives include Patient-Centric EHR, Precision Medicine, Population Health, Learning Health, Big Data, EHR-based pragmatic clinical trials, International Patient Summary, and Health Information Exchange (HIE). The acceptance and implementation of the standards and the applications they enable depend on the use case and the

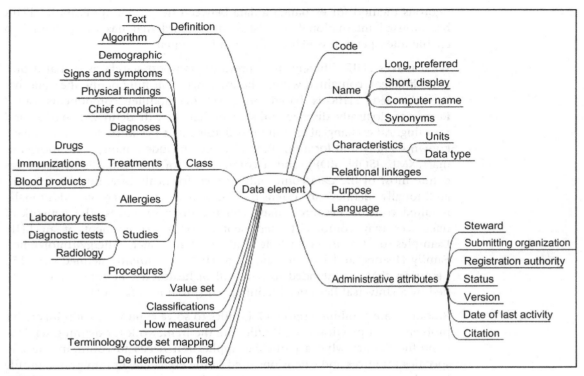

**FIGURE 4.1**
Data element showing a full set of attributes.

perceived value of the application. In the simplest of terms, these initiatives all require the exchange of data between and among disparate sites, and we want to understand and use the data exchanged. Success requires a predefined trigger and content that is of interest to the receiver. The exchange of data may be a push process with the content predefined or a query, in which the content is requested. In the United States, HIEs have not met with sustainable success for the most part because of a mismatch between what is sent and what is required.

## Fast Healthcare Interoperable Resources (FHIR)

The latest in an evolution of transport standards from HL7 is FHIR (https://www.hl7.org/fhir/, 2016). FHIR focuses on web-based technology, simplicity, comprehensibility, ease to learn, ease of implementation, and flexibility. FHIR reflects the lessons learned from previous HL7 standards: v2 series, v3 messaging and related products such as CDA and CCD. FHIR provides a rigorous method for exchanging data between healthcare applications. FHIR has aroused international interest as a standard that has the potential to enable interoperability. FHIR is free and fully open.

What makes FHIR different from previous transport standards is that it provides linkages, contained within the exchange, that fully define the data that is exchanged. FHIR is based on a set of modular components called Resources, logically discrete units of exchange with defined behavior and meaning. All exchangeable content is defined as a Resource. Resources have a known identity and location. Resources also include a name, definition, format (XML, JSON, RDF), and a common set of metadata. FHIR Resources define most of the data commonly needed for healthcare, but the standard additionally defines an extension mechanism to cover an additionally required content. Resources may reference other resources. The network of references may combine to represent a specific component of an EHR. Examples of Resources include Patient, Practitioner, Allergy Intolerance, Family History, and Patient Care Plan. HL7 has currently defined over 150 Resources that are intended to cover all of healthcare. Resources are identified by a Universal Resource Identifier (URI) that is in fact a URL.

Resources are combined into HL7 Profiles to solve clinical and administrative problems in a practical way. Profiles are the framework for defining services. A profile defines what a particular application needs to communicate and specifies resources and extensions used in the exchange. Examples of a profile include referral of a patient to another facility, supporting a HIE, populating and maintaining a registry, ordering a medication, exchanging genomic data, or providing data to a clinical decision support algorithm such as a risk assessment calculation.

FHIR uses a RESTful API to exchange Resources between systems. Representational State Transfer is a software architectural style for how to connect systems consisting of guidelines and best practices for creating scalable web services. RESTful systems typically communicate over HTTP verbs (GET, POST, PUT, and DELETE). Data may be exchanged as a "bundle" of resources that may be messages or documents. FHIR is service-driven, which means you can send just the data that is required for a specific purpose.

The Office of the National Coordinator for Health Information Technology (ONC-HIT) has initiated the creation of two FHIR-based Information Guides. The first, Structured Data Capture (SDC), defines a container for pushing information to a receiver. The second, Data Access Framework (DAF), defines a process through which data may be requested from a sender. Both have the potential to be of value in the global community. Example of the use of SDC is reporting data to Public Health, submission of medical documentation for various purposes, or populating a cancer registry, or populating a rare disease registry.

## IHE Standards

IHE has created a set of profiles (IHE Profiles) that expands interoperability across networks by identifying how existing standards (HL7, DICOM, IETF, OASIS, ISO, etc.) may be used for end-to-end communication and exchange of data. IHE Profiles serve to extend siloed interoperability across silos. Some of the commonly used IHE Profiles are:

- XDM—cross-enterprise Document Media Interchange. XDM transfers documents and metadata via physical media such as CDs, USB removable media, and email attachments. XDS manages patient identification, document identification, descriptions, and relationships. This profile is particularly important for transporting images.
- XDR—cross-enterprise Document Reliable Interchange. XDR supports the exchange of documents between healthcare sites using a web-based, push communication without the need for a document repository. XDR is useful for populating registries or supporting a patient-centric EHR where patient receives care at multiple sites.
- XDS—cross-enterprise Document Sharing. XDS supports the sharing of documents between any healthcare enterprise through a common registry. Documents are registered and stored in the registry. There they may be found and accessed. A variation of this Profile also supports images. XDS supports global sharing, with patient consent, of a Patient Care Summary.
- PDQ—Patients Demographics Query. PDQ establishes a central patient information server based on patient demographic information.
- PIX—Patient Identifier Cross Referencing. PIX is used for cross-referencing multiple patient identifiers across disparate systems using different identifiers.

These IHE Profile standards are important in global health, particularly when the focus of global health is to enable cross-border communications. Knowing where to look for needed data and having access to patient identification, patient consent, and patient data is a necessity for cross-border sharing.

## ISO/JIC Standards

ISO TC 215 has created a wealth of standards over the years plus incorporating standards from other SDOs. One product that has value to the global community is the five standards for the Identification of Medicinal Products (IDMP) (http://www.ema.europa.eu/ema/index.jsp?curl = pages/regulation/general/general_content_000645.jsp, 2016). This standard was create under the auspices of the JIC. The purpose of IDMP is to simplify the exchange of data among all interested stakeholders. The IDMP defines the data elements and structures for the unique identification and exchange of regulated information on:

- Substances (ISO 11238)
- Pharmaceutical dose forms, units of prescription, routes of administration, and packing (ISO 11239)
- Units of measurement (ISO 11240)
- Regulated pharmaceutical product information (ISO 11616)
- Regulated medicinal product information (ISO 11615)

HL7 has created a standard called the Common Product Model for the exchange of medicinal data that is compliant with IDMP. The EMA is implementing IDMP within the European Union (EU) for the exchange of medicinal products data.

## JIC Project

JIC has an ongoing project to create a Patient Summary Standards Set (http://www.jointinitiativecouncil.org/news/JIC_Patient_Summary_Standard_Set-Foundation%20_Scope_Report_20151008_v3_5.pdf, 2016) to provide for the transport of patient summary data both within a country and across country borders. The work is being done by four subgroups: use case development; standards identifications and analysis; implementation & guidance document development; and conformity assessment. The content of the Patient Summary will be defined. The success of this project will be to define real use cases that have value and engagement of the healthcare systems in all countries.

## LOOKING TO THE FUTURE

The current state of the world relative to standards and interoperability is the mix between "no standards" and "too many standards." Interoperability at best exists for a particular use case. Approximately 80% of the work currently done in this area is to produce "work-arounds" unsolved issues for which we know a solution but have not made the commitment to solve.

OpenMedicine has a project to create a master database for identification of medicinal products based on IDMP. The model is based on substance, pharmaceutical product, medicinal product, and medicinal product package. This master data set will be the basis for the exchange of medicinal information between countries. If a person from Country A is in Country B and requires a medication, then Country map will map its own identification of the drug to the master set. Then Country B will map from the master set to its own identification. This solution is to be applauded because it will solve an important problem. There are many use cases which this project will solve. But this is a work-around solution. The best solution would be for all countries in the world to adopt the master database for use within each country of the world. The cost savings would be tremendous. Medical errors would be reduced. If, in addition, a unique number would be assigned to each medicinal product at the beginning that would last throughout its lifetime, a prefix could be added that would identify where in its life cycle the product existed. This number could change as the product moved through the life cycle.

In the United States, queries and the exchange of data among sites requires mapping from each local database into a master set, always with some loss of information. If we could globally adopt a single master database for data representation throughout healthcare, integrating data would be easier, more accurate, and timelier. It would be a single fix and would cost money. Even so, it would be less expensive that to continue mapping from one representation of data to another.

Changes such as this can only occur when all the stakeholders come together and push for the best solution. It would require a partnership to a common cause between the public and private sectors.

To address the gaps in standards and even some of the duplication of standards, the future provides an opportunity to reverse the process of trying to do something and selecting the standards, if they exist, to do it. The future needs vision, innovation, and disruption to remove the constraints of laws, policy, existing processes and procedures, unwillingness to change, and competition over collaboration. There are too many challenges and unsolved problems to reinvent the wheel. We must divide the work and enjoy the result of mutual solutions. That is what global health is all about.

## References

Shortliffe, E., Cimino, J., 2012. Biomedical Informatics. Springer, London.

Richesson, R., Andrews, J., 2006. Clinical Research Informatics. Springer, New York, NY.

<http://dicom.nema.org> (accessed 15.07.16).

<http://www.astm.org/COMMITTEE/E31.htm> (accessed 15.07.16).

<http://www.hl7.org> (accessed 15.07.16).

<http://www.ieee.org> (accessed 15.07.16).

<http://www.ihe.net> (accessed 15.07.16).

<http://www.cdisc.org> (accessed 15.07.16).

<http://www.cen.eu> (accessed 15.07.16).

<http://www.iso.org/iso/iso_technical_committee?commid=54960> (accessed 15.07.16).

<http://www.openEHR.org> (accessed 15.07.16).

<http://www.jointinitiativecouncil.org> (accessed 15.07.16).

<http://www.gs1.org> (accessed 15.07.16).

<http://www.who.int/whosis/icd10/> (accessed 15.07.16).

<http://www.ihtsdo.org> (accessed 15.07.16).

<https://en.wikipedia.org/wiki/ICPC-2_PLUS> (accessed 15.07.16).

<http://loinc.org> (accessed 15.07.16).

<https://www.nlm.nih.gov/research/umls/rxnorm/> (accessed 15.07.16).

<http://www.va.gov/TRM/StandardPage.asp?tid = 5221^> (accessed 15.07.16).

<https://en.wikipedia.org/wiki/ISO/IEC_11179> (accessed 15.07.16).

<https://www.hl7.org/fhir/> (accessed 15.07.16).

<http://www.ema.europa.eu/ema/index.jsp?curl = pages/regulation/general/general_content_000645.
jsp> (accessed 15.07.16).

<http://www.jointinitiativecouncil.org/news/JIC_Patient_Summary_Standard_Set-Foundation%
20_Scope_Report_20151008_v3_5.pdf> (accessed 15.07.16).

# CHAPTER 5

# Data Systems and Global Health Informatics Research

**J.J. Cimino[1], E. Ayres[2] and R. Pillay[1]**
[1]University of Alabama at Birmingham, Birmingham, AL, United States
[2]NIH Clinical Center, Bethesda, MD, United States

## BARRIERS TO GLOBAL SHARING OF HEALTH DATA

To learn from data, we need very large datasets, larger than any entity can collect on its own. Fortunately, we are living at a watershed moment in history, where it is becoming possible to access large numbers of data points in individuals, along with clinical information. In principle, the wealth of these data should accelerate progress in biomedicine—making it possible to integrate research and clinical information to understand disease causation, response to treatment, and new treatment opportunities. In practice, however, we are not organized to seize this extraordinary opportunity—nor are we on a path to do so. For the most part, data are collected and studied in silos: by disease, by institution, and by country, resulting in sample sizes inadequate to make robust discoveries. If we remain on the current path, the likely outcome will be an assortment of fragmented systems that inhibit the acceleration of medical progress.

Regulatory and ethical procedures could not anticipate the exponential advances in technology and thus were not designed to enable widespread comparison across studies and the sharing of information across datasets, institutions, and countries. In the absence of an open and interoperable solution, the current closed, proprietary systems create a fundamental barrier to gaining the benefits of data aggregation and thereby slow the understanding, diagnosis, and treatment of disease—a case in point is the fragmented medical records system of the United States which has for decades inhibited the quality of patient care and the ability to learn from experience.

## CONTENTS

**109**

Many segments of the global biomedical ecosystem are negatively affected by the inability to share data in a safe, secure, and reproducible manner:

Governments and Foundations increasingly require the data they fund to be made broadly available. However, they lack mechanisms to manage sharing of data, and much of the data are unused.

Biopharma relies on information both to identify new targets for therapy, and for patient stratification and the design of clinical trials. However, companies find it difficult to obtain access to genetic and clinical information, lack internal expertise and infrastructure to perform analyses, and worry about liabilities associated with data regarding informed consent and privacy.

Scientific Researchers collect information at a prodigious rate but generally lack access to software tools and computational infrastructure needed to manage this magnitude of data. Investigators working in each disease solve the same set of computational and software challenges, but often do so in ways that are not interoperable. Few have access to sample sizes needed to achieve power, and thus to forge unexpected connections among diseases.

Hospitals and healthcare systems increasingly need to collect, store, and interpret information, but it is expensive for each to individually create software tools and infrastructure. On their own, each lacks the critical mass of comparative data required to care for their own patients. The lack of norms for sharing and protecting data are slowing the uptake of science in the clinical setting.

Advancing medical knowledge and improving clinical care will require the widespread ability to access scientific and clinical data in a secure and trusted manner, and to enable comparisons thereof. As regulations and attitudes toward sharing data vary within and across national boundaries, and as different sectors have different needs and goals, it is necessary to enable interoperability while preserving diversity of approach and application. To achieve these goals, an appropriate structure, technology, and regulatory and ethical framework is needed to bring together researchers, healthcare providers, funders, disease advocacy groups, life science and technology companies, and informed citizens to enable, support, and promote the responsible sharing of data.

In addition to social and regulatory issues, significant technological barriers must be overcome in order for sharing of data to positively impact global health. Clinical data repositories must be developed, which are able to integrate data from multiple sources, using multiple terminologies, for multiple uses by multiple constituencies. We describe one such repository, created at the US National Institutes of Health, which illustrates some successful approaches to address these challenges.

# THE US NATIONAL INSTITUTES OF HEALTH'S CLINICAL DATA REPOSITORY

The Biomedical Translational Research information System (BTRIS) serves as the clinical research data repository for the National Institutes of Health (NIH) intramural program (Niland et al., 2006). Serving as a research tool for the NIH clinical research community, BTRIS maintains multiple applications to support active protocol management, hypothesis testing, mandatory reporting, and the provision of data to institute clinical trial management systems. BTRIS stores in perpetuity data from the NIH Clinical Center electronic health record (EHR) system (the Clinical Research Information System (CRIS)) as well as data from multiple institute clinical trial management systems, clinical support systems (such as pulmonary function tests), institute databases, and individual researchers.

## Data Content Requirements

*Data Sources.* The NIH is located on a 1.2-km$^2$ campus in Bethesda, Maryland, just north of Washington, DC. All 27 of the NIH's Institutes and Centers (ICs) maintain a presence on the campus, and most of them house intramural programs that conduct clinical research at the Clinical Center, a 240-bed hospital on the north end of campus. Similar to most hospitals in the United States, the Clinical Center uses a commercial comprehensive EHR system. Locally referred to as the CRIS, it is a commercial system (Allscripts Solutions, Chicago, Illinois) that has been in place since 2004. Unlike most hospitals, the Clinical Center had a previous EHR dating back to 1976, called the Medical Information System (MIS), sold by Technicon Data Systems (Oakland, CA). When CRIS was installed in 2004, all MIS data were archived. Despite its name, CRIS is primarily a patient care system, although since all patients seen at the Clinical Center are research patients, all the data in CRIS are technically research data. Thus they are used for both care of the individual patient and to address primary research questions that are the subject of NIH studies.

Several other NIH ICs maintain their own clinical information systems, primarily in the form of clinical trials data management systems (Cimino et al., 2014). These range from commercial products, such as LabMatrix (Biofortis, Columbia, MD), to homegrown systems, such as the National Institute of Child Health and Development's Clinical Trials Database (CTDB), as well as ancillary systems in the Clinical Center. At last count, there were 19 such systems, collecting data ranging from laboratory reports from non-NIH referral laboratories, to genetic sequence data, to case report forms. Table 5.1 lists the current systems contributing data to BTRIS. These systems include a wide

**Table 5.1** BTRIS Data Sources

| System Name | Institute | System Type | Developer | Data Domains |
|---|---|---|---|---|
| Medical Information System (MIS) | Clinical Center | EHR | Commercial | Demographics, Laboratory Tests, Blood Bank, Medications, Microbiology, Radiology, Vital signs, *Clinical Notes, Anatomic Pathology* |
| Clinical Research Information System (CRIS) | Clinical Center | EHR | Commercial | Demographics, Subject-Study Attribution, Vital Signs, Clinical Documentation, Alerts, Allergies, Observations, Document Images, Medication Administration, Medication Orders, *Other Orders*, *Admission/ Discharge/Transfer* |
| Softlab | Clinical Center | Ancillary Department System | Commercial | Clinical Laboratory Tests, Microbiology Tests, Anatomic Pathology, Blood Bank Tests, Blood Bank Products |
| Softmed | Clinical Center | Medical Records Department System | Commercial | Admission Notes, Discharge Notes, Diagnoses, *Other Dictated Notes* |
| Vmax | Clinical Center, NHLBI | Ancillary Department System | Commercial | *Pulmonary Function Reports* |
| Jaeger | Clinical Center | Ancillary Department System | Commercial | *Pulmonary Function Reports* |
| Pain and palliative Care System | Clinical Center | Ancillary Department System | Commercial | *Pulmonary Function Reports* |
| LinkTools | Clinical Center, NHLBI | Ancillary Department System | Commercial | Electrocardiograms |
| ProSolv | Clinical Center, NHLBI | Ancillary Department System | Commercial | Echocardiology Reports |
| RadNet | Clinical Center | Ancillary Department System | Commercial | Radiology Reports |
| Carestream | Clinical Center | Picture Archiving and Communication System | Commercial | Radiographic Images |
| Protrak | Clinical Center | Protocol Services Department System | Clinical Center | Studies, Investigators |
| Clinical Research Information Management System of the NIAID (CRIMSON) | NIAID | CTDMS | NIAID | Study-Subject Attribution, Laboratory Tests, Medications, patient Problems |

*Continued...*

**Table 5.1** BTRIS Data Sources *Continued*

| System Name | Institute | System Type | Developer | Data Domains |
|---|---|---|---|---|
| Clinical Research Database (CRDB) | NIAAA | CTDMS | NIAAA | Assessments (Surveys) |
| Labmatrix | NCI | CTDMS | Commercial | Biospecimens |
| Cancer Central Clinical Database (C3D) | NCI | CTDMS | NCI | Study Attribution, Laboratory Tests, Case Report Forms |
| Clinical Trials Database (CTDB) | NICHD, NIAAA, NIDDK, Clinical Center | CTDMS | NICHD | Encounter Forms |
| Labmatrix | NHGRI | CTDMS | Commercial | *Biospecimens, Case Report Forms* |
| Varsifer | NHGRI, NIMH | Laboratory Database | NHGRI | Whole Exome Sequences |

variety of data, including subject demographic data, outside laboratory reports, text documents, images, biospecimen information, and genetic sequencing. Most of these data are associated with a Clinical Center patient, allowing them to be pooled in BTRIS, such that all data captured regarding a single subject are associated via their medical record number. In other cases system-specific unique identifiers are used to aggregate data on individual subjects from that system within BTRIS.

### Tracking Clinical Protocol Enrollment

As mentioned earlier, each patient seen at the Clinical Center is a subject enrolled in one or more research studies, referred to at NIH as *protocols*. The Protrak system (see Table 5.1) provides information on the subjects enrolled in each protocol, including medical record number, date of enrollment (consent date), and, if relevant, date of withdrawal from the protocol. Protrak also provides information on the investigators associated with the study, including the Principal Investigator (PI) and other researchers who are authorized to view study data.

### Retrieving Data Based on Protocol

Most of the systems contributing data to BTRIS have precise tagging of those data with respect to the relevant research protocol. Older systems such as MIS are lacking these data, while only some data in BTRIS are so tagged. In particular, users of CRIS are instructed to select a protocol from a pull-down list when entering orders (e.g., laboratory tests or medications). Unfortunately, this approach to tagging is not always accurate and, even when it is correct, it fails to account for situations where a test or treatment is relevant to more than one protocol for the same patient (ordering the same test multiple times, once for each protocol, is not desirable for many reasons). Therefore BTRIS never assumes that a particular datum is relevant to any particular one of the subject's protocols. Rather, it assumes that any datum may be attributable to a protocol if its occurrence falls on a date that is between the subject's enrollment data on that study and either the termination of the study or the date that the subject withdraws from the study (an unusual situation). Although this adds a layer of complexity to retrievals, it allows a broad enough retrieval that no relevant data will be missed but avoids unnecessary retrieval of data known not to be relevant (i.e., outside the date range of the study).

### Deidentified Data

Although almost all data coming to BTRIS are associated with a known research subject, and there are header fields that contain identifiers, many of the fields in the data do not contain identifiers (e.g., the "Result" field of a laboratory test). Some text fields, such as the radiologist's interpretation of an X-ray report, do not typically contain identifiers, whereas other text fields, such as text of a hospital admission note, can almost be guaranteed to contain identifying information. These identifiers are not problematic when data are being accessed as part of an Institutional Review Boards (IRB)-approved study because subjects of these studies have typically given consent (or the IRB has authorized a waiver of consent) that allows them to be seen by the researchers involved in the study.

BTRIS does not provide identifying information to users who are not authorized to see data from the subject's protocol. However, they are permitted to see data in deidentified form. In this case retrieval will be restricted to those fields known to be free of identifiers. For text fields that may have identifiers, a corresponding deidentified field is defined, the text is deidentified with commercial software, and then the resulting redacted text is stored in the database along with any other results. Deidentified queries are then restricted to those fields (such as numeric test results, temperature, and deidentified text) that are known to be identifier-free.

## Data Model and Data Base Design

### Overall Data Model

Fig. 5.1 shows the overall BTRIS data model, which can be considered in several sections. The tables with the white background (left side and upper left corner) represent information relating to and among study protocols, subjects, investigators, and users in a classic entity—relation diagram. The longest table in this section is the Subject table and includes a wide variety of data that will only exist once for each subject (e.g., date of birth, home address, etc.). The three upper-right groups of tables with shaded backgrounds contain data related to the data storage activities. The shaded group of tables at

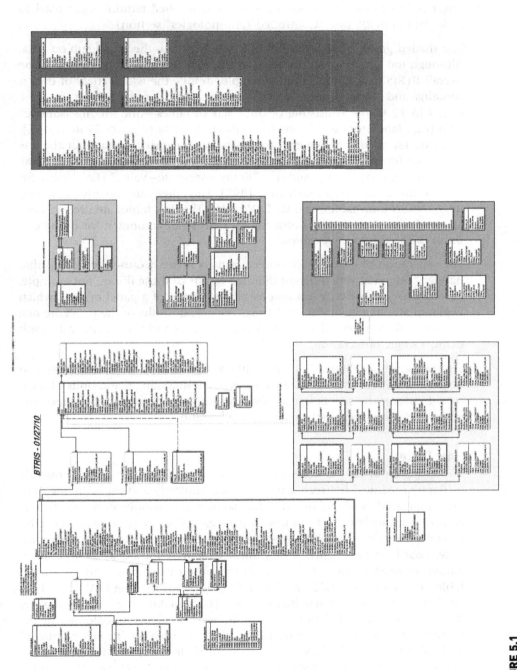

**FIGURE 5.1**
BTRIS data model.

lower right contain information about the controlled terminologies used to code data in BTRIS (see "Controlled Terminologies" section).

The shaded group of tables at bottom center contain the actual subject data. Although too small to read in detail in this figure, it should be clear that the overall BTRIS data model is rather simple, despite the wide variety of data it contains and the functions it performs. In particular the subject data are stored in 12 tables, consisting of three sets of tables—one for "measurable" data (e.g., laboratory tests), one for "substance" data (e.g., medication data), and one for all other "general" data (e.g., text reports). In each set there is one table for the "event" (what happened), one for the "observation" (the result of what happened), and one "Entity–Attribute–Value" (EAV) table for each of the main tables (Nadkarni, 1997). Thus there are three Event tables, three Observation tables, and six EAV tables. The EAV tables are used to store any attribute of an event or observation that is not accounted for in the columns of those respective tables.

The separation of events from observations is made because the "things that happen" are often very different than the results of those things. For example, an order for a laboratory test involves the selection of a panel of tests, which is placed at a particular date and time. The actual results of the panel are one or more values, reported at a somewhat later date and time, with each result being a single observation.

The Measurable Observation table differs significantly from the other two observation tables by including columns for normal ranges and units of measure. However the separation of observations and events into the three categories was primarily a performance consideration.

### EAV Tables

The inclusion of EAV tables for each of the six main tables serves to provide a "happy medium" between a model that completely and explicitly incorporates that models of the original data sources and simply stores all data as entities and attributes of entities in a single table, as is done in the i2b2 model (Weber et al., 2009). A subjective decision was made for each attribute, based on experience with clinical data warehouses, as to whether it would be retrieved frequently enough to warrant its own column in a main table, or whether it would be relegated to the corresponding EAV table. Thus attributes such as "Normal Range" and "Units of Measure," which are frequently reported with laboratory results, were modeled as columns in the Observation Measureable table, while attributes such as "Reporting Status" and "Signature" are in EAV tables. Each attribute is represented as a coded entity, included in the Research Entities Dictionary (RED; see "Controlled Terminologies" section).

## Time

Temporal elements in clinical data are often complex. For example, a laboratory result may be reported with the time the test was ordered, the time the specimen was obtained, the time the specimen arrived in the laboratory, the time the test was performed, the time the result was reported, or any combination of these. In BTRIS we attempt to identify the most relevant time for a particular datum. For example, for a laboratory test, the time the specimen was obtained is the most clinically relevant, followed by the time the specimen was received in the laboratory, and so on. In the past NIH laboratory results had, at best, the time of arrival in the laboratory. More recently, however, phlebotomists using barcode readers are able to capture the time the specimen is obtained from the patient. The Time columns in the BTRIS tables include the "best" time available, with all times (including the best one) stored in the corresponding EAV table. This allows the best time to be easily retrieved, but all times are available if desired. In addition, because all the time attributes are coded in the RED and included in a single class, it is easy to retrieve all times for a particular observation with a single query.

## Controlled Terminologies

### Ontology Structure

NIH data generally include some coding using controlled terminologies. With the exception of diagnoses coded with the International Classificaiton of Diseases (ICD), most coding systems used at NIH are local terminologies. Our approach to managing these terminologies follows the approach taken with the Medical Entities Dictionary at Columbia University (Cimino, 2000). Each term in each source terminology is mapped to a concept in the our enterprise ontology, called the RED (Cimino et al., 2014). Where an external term is synonymous with an existing concept in the RED, the term is mapped as a synonym of the concept. In most cases, however, the term will be considered to have its own meaning, separate from other similar terms in other terminologies, and will be given its own unique identifier (RED Code).

Concepts in the RED are organized into a semantic network that includes is-a relationships in a directed acyclic graph. All concepts (except the top node "Research Entity") have at least one parent term and may have more than one (i.e., the RED is a polyhierarchy). Concepts may have literal-value attributes (such as their names and codes from the source terminology, synonyms, etc.) and may have attributes that relate them to other concepts through semantic relationships (i.e., the attributes have RED Codes as values). The semantic network allows inheritance of semantic relationships, where appropriate (see Fig. 5.2).

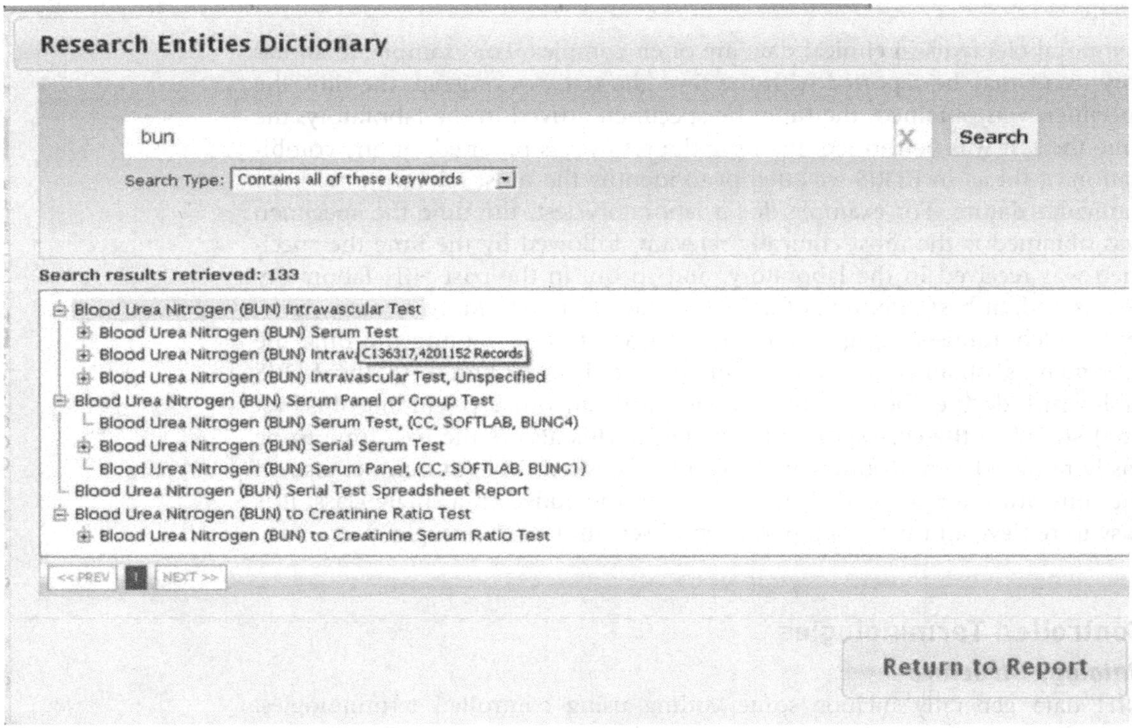

**FIGURE 5.2**
The NIH Research Entities Dictionary.

### Ontology Maintenance

The RED is maintained with Terminology Development Environment (TDE), a multiuser commercial ontology management tool from Apelon (Hartford, CT). The NIH employs two full-time ontologists (a pathologist and a pharmacist) to maintain the RED, adding new terminologies, adding new terms and concepts as existing terminologies change, and organizing the RED hierarchy into classes that are useful for retrieval and display of data.

### Storage of Data Coded With Controlled Terminologies

When data are stored in BTRIS, the extraction-transformation-loading (ETL) process includes a RED-encoding step that uses a source-specific translation table to replace the local term codes with RED codes. Each translation table includes a column for the source code and a column for the RED code; the replacement is straightforward. Retrieval of data is similarly straightforward. When a user wishes to retrieve a particular type of data—say, e.g., a particular type of laboratory test result, the user selects a name from a list of available terms and the system creates a query that uses the RED code associated with the term.

### Retrieval of Data Coded With Controlled Terminologies

If the user requires a class of data (e.g., all serum glucose results), the system makes use of a special table called the "Ancestor-Descendant-Identity" table. This (ADI) table has two columns, Ancestor and Descendant, and has a row for each concept and each of its ancestors. Thus, if "Serum Glucose" has a child term in the RED called "Serum Glucose by Mass Concentration" (i.e., reported in milligrams per deciliter) which in turn has two children, say "Serum Glucose 1" and "Serum Glucose 2," then there will be a row in the ADI table for each test and its parent class (Serum Glucose by Mass Concentration), each test and its grandparent class (Serum Glucose), and one for the parent–child relationship between Serum Glucose and Serum Glucose by Mass Concentration. There is also one row for each of these four concepts, in which the RED codes are in both the Ancestor and Descendant columns (hence the inclusion of "Identity" in the name). In effect, each concept is considered to be a member of its own class.

Retrieving data with the ADI tables require a "join" between the ADI and the table with the desired data, in which the concept of interest is matched to the Ancestor column and all rows in the target table for which the value in the column of interest matches one of the Descendant values from the rows selected in the ADI table. This query has proven to be extremely efficient for supporting class-based queries. It also simplifies query design. Because of the inclusion of the "identity" rows, the query developer does not have to worry about whether the data are stored with a particular RED code or a descendant of the RED code—the query simply always includes the join with the ADI table.

### Case Study: ICD9CM and ICD10CM

The power of this approach, with respect to both terminology maintenance issues and merging data across multiple sources, can be demonstrated by considering queries for patient diseases. Prior to 2015, patient diagnoses in US hospitals were coded with the 9th edition of the International Classification of Diseases with Clinical Modifications (ICD9CM). Since 2015, these data have been coded with the 10th edition (ICD10CM), an almost completely different set of terms and codes. Prior to this change, all ICD10CM codes were added to the RED, with about 3000 terms mapped to existing RED concepts drawn from ICD9CM. The remaining terms were added as new concepts, inserted into the existing ICD9CM hierarchy. The ETL process was able to include codes from ICD9CM and ICD10CM in a single translation table, since there were not overlaps in the codes. This allowed the data loading process to continue uninterrupted as the Clinical Center transitioned from ICD9CM to ICD10CM. Querying for the data was also unaffected: a user selecting a concept of interest would use the ADI table to obtain all terms in a particular class, regardless of whether they were coded in ICD9CM or ICD10CM (Cimino and Remennick, 2014).

## Data Access Authorization

### Retrieval of Identified Data

A principle use of BTRIS is the retrieval of data relevant to research study, across patients (i.e., all human subjects in the study) and across all data sources (e.g., CRIS, MIS, and the system used by the institute conducting the study). For this purpose, the query makes use of a set of patient unique identifiers associated with the study (or a user-specified subset of those identifiers), and the time period during which each patient was a enrolled as a subject in the study (see "Data Content Requirements" section). Access to these data (in identified form) is restricted to the PI on the study and individuals specified by the PI, such as other investigators who are participating in the study and data managers.

### Retrieval of Deidentified Data

All data in BTRIS have been collected as part of some NIH-sponsored study and, by federal regulations, made available to other investigators in deidentified form for reuse for research purposes. Deidentification refers to the removal of information that could uniquely identify an individual. In addition to direct identifiers, such as names and tax identification numbers, indirect identifiers such as addresses and phone numbers must be removed. While information such as postal codes and dates of events could be used as a key to reidentification, studies have shown that their removal only interferes with the usefulness of the data but does not hinder determined reidentification efforts (Cimino, 2012) and so their inclusion is allowed in what are referred to as "limited use datasets" (Dokholyan et al., 2009). Although IRBs consider limited use datasets to constitute information on human subjects, they are exempt from IRB oversight if the purpose of the usage is documented and access to the data is restricted to investigators directly involved in the study.

From a practical standpoint, deidentification entails restricting data reporting to specific fields of data that are known to be identifier-free, such as the name and result of a laboratory test or the text report of an electrocardiogram. Fields such as name and medical record number are obviously excluded. In the case where text fields may include identifiers, such as admission notes and the like, a second version of the field is created with identifiers stripped out using a deidentification software.

## *Notification of Data Reuse*

Although IRBs impose restrictions on reuse of data to protect of patient privacy, there are not legal protections for the original investigator. Investigators can control the data within their own data management systems, but data collected in an EHR in a clinical setting are open to other users. This leaves the original researcher vulnerable to having a second investigator—possibly without knowledge of the first investigator's work—to obtain, analyze, and publish the first investigator's data. For example, a researcher could seek laboratory results on patients taking a particular drug to examine adverse effects on the liver or kidneys. This scenario is a very real possibility when using clinical data repositories containing EHR data from medical centers conducting clinical research.

Because all data in BTRIS are associated with clinical studies, it is possible to determine which studies are responsible for the origin of individual data retrieved in a deidentified query. Although there is not absolute restriction on reuse, the NIH has established a policy that provides recognition of the original investigator's interests in controlling such reuse. In general, reusers of data must obtain approval from original investigators to reuse data if the data significantly overlap with the original study and the research question significantly replicates the original purpose of the study.

All data retrieved from BTRIS in deidentified queries are tagged with information about the original study and investigator. This information is provided so that the BTRIS user can learn more about the source of the data, if desired, and possibly propose a collaboration with the original investigator. Some data are also flagged as requiring permission for reuse. Data can be reused without permission if any of the following are true:

- The data are not known to be associated with a study (this occurs with older data, prior to collection of enrollment data).
- The study has been terminated for more than 2 years.
- The dataset does not include more than 1% of the subjects from the study.
- The study constitutes less than 10% of the data in the dataset.

A typical scenario might be a query for all blood pressure and blood glucose results to look for some correlation. This will return many millions of results on hundreds of thousands of patients. While many of the results will be from patients who are subjects of active studies, and may include up to 100% of subjects from a particular study, not one study will comprise 10% of the total dataset, so no permissions are needed.

If all of the above are true, then when the data are downloaded, the original investigator is notified of the access and told that the BTRIS user will be in contact to obtain permission for reuse. The BTRIS user is told that the original investigator has been notified and must be contacted for permission. If the user makes use of the data (e.g., uses it as preliminary data for a grant proposal or publishes it in a journal paper), this will be considered a breach of the BTRIS data use agreement. If the user contacts the investigator and the investigator does not grant permission for use, the user may appeal to the BTRIS Data Use Committee, who may determine that the purpose of the reuse does not significantly overlap with the original research question.

## User Interface Components

Queries for identified data are accomplished through a self-service user interface developed with the business intelligence software tool Cognos (IBM, Armonk, NY). The user completes a query template that is specific for each data type (laboratory test results, medication data, diagnoses, etc.), such as the one shown in Fig. 5.3. The user must first selects the research study or studies of interest and may either run the report at that point (obtaining all data of the relevant type for all subjects in the seleted studies) or may limit the report to select data for certain subjects, limited to specified dates, and limited to particular data types, based on the selections from the controlled terminology (see "Terminology Lookup in BTRIS" section). When the user clicks on the Run Report button, the results are returned (usually in a few seconds) in a tabular form, which can be downloaded as a tab-delimited data file suitable for loading into a spreadsheet or other data analysis software (Fig. 5.4).

Deidentified data queries typically use selection criteria that cross over data domain boundaries (e.g., "find me all the laboratory tests on patients with a particular disease and race"). We found this task to be too complex for development in Cognos, which makes use of predefined, fixed templates. We therefore built a customize system that allows users to construct complex queries using filters selected dynamically. Filters are dragged from the list on the left ("Diagnosis," "Demographics," "Medications," and "Lab[oratory] Tests"—see Fig. 5.5), one at a time, to create a vertical list of selection filter frames. Users who fill out parameters in each frame as desired and then link them with logical relationships ("and," "or," and "not"). Fig. 5.5 shows an example of the user interface, with a query that has been performed to find all diagnoses on all patients in the database with an elevated result on a vitamin B12 level. Where filter parameters use controlled terminologies (such as the names of laboratory tests), the user interacts with the controlled terminology (see "Terminology Lookup in BTRIS" section).

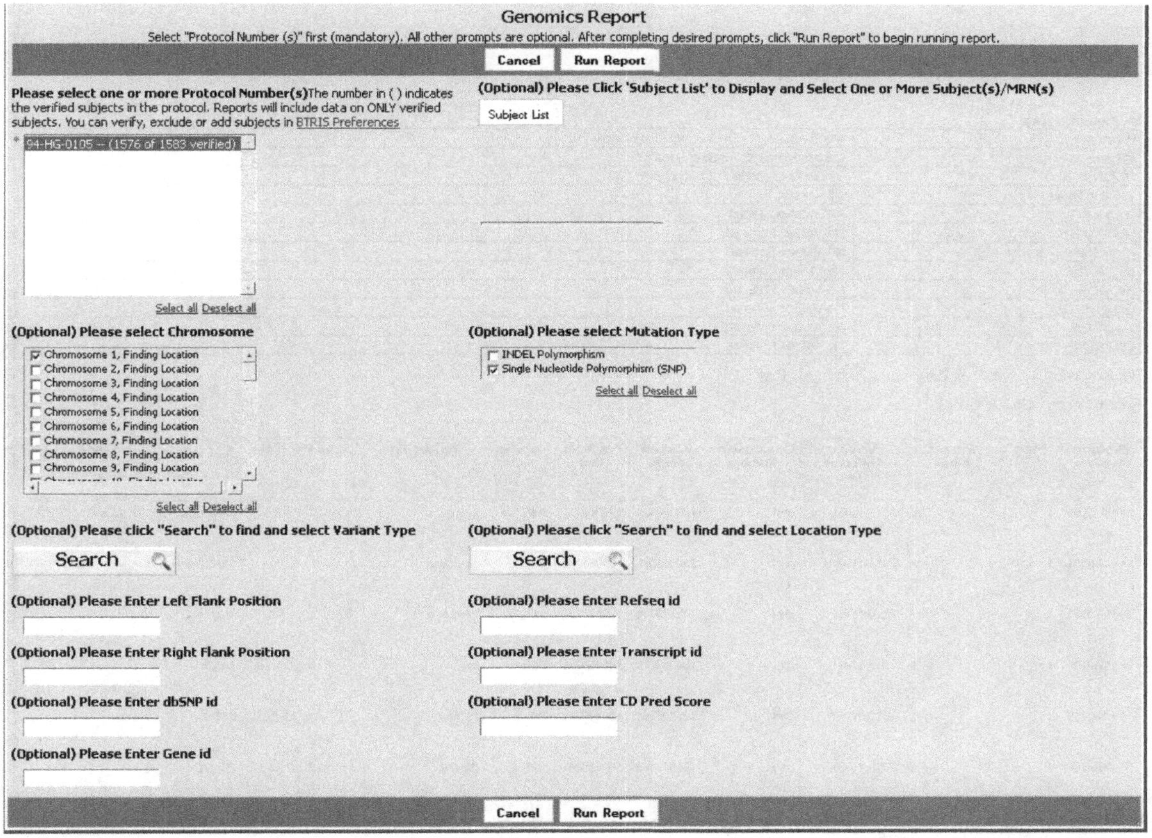

**FIGURE 5.3**
BTRIS identified data query template.

## Terminology Lookup in BTRIS

One advantage of the BTRIS query tools is that they make use of the RED, described in "Controlled Terminologies" section. The RED lookup tool allows the user to search the RED for terms based on their names and synonyms and to search through the hierarchy to specify the level of detail desired. For example, if a user is performing an identified query to retrieve data on subjects from a study where a very specific vitamin B12 level is performed (e.g., by a special chemistry laboratory), the user enter "B12," and is shown (perhaps among other things) the portion of the RED hierarchy that contains vitamin B12 laboratory test terms. The user can then navigate down the hierarchy until she finds the specific test of interest. On the other hand, a user performing a deidentified query may not care about all the different test

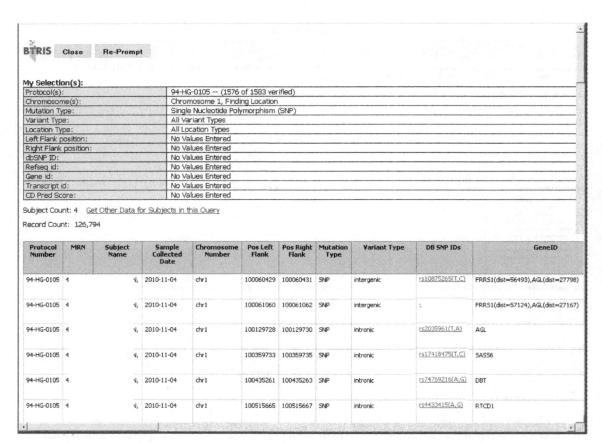

**FIGURE 5.4**
BTRIS identified query—Results.

terms. In this case he can simply select the RED term *Vitamin B12 Intravenous Test* ("intravenous" refers to tests performed on serum, plasma, and whole blood) to obtain all results for all tests in that class.

## BTRIS and Global Health

The NIH Clinical Center and the NIH intramural program support early-stage clinical trials and a focus on the study of rare diseases. The international community offers the opportunity to study unique subject cohorts through the participation of international subjects participating in clinical trials at the Bethesda, Maryland campus. Since the inception of EHRs at the NIH Clinical Center in 1976 to the present, the BTRIS repository has

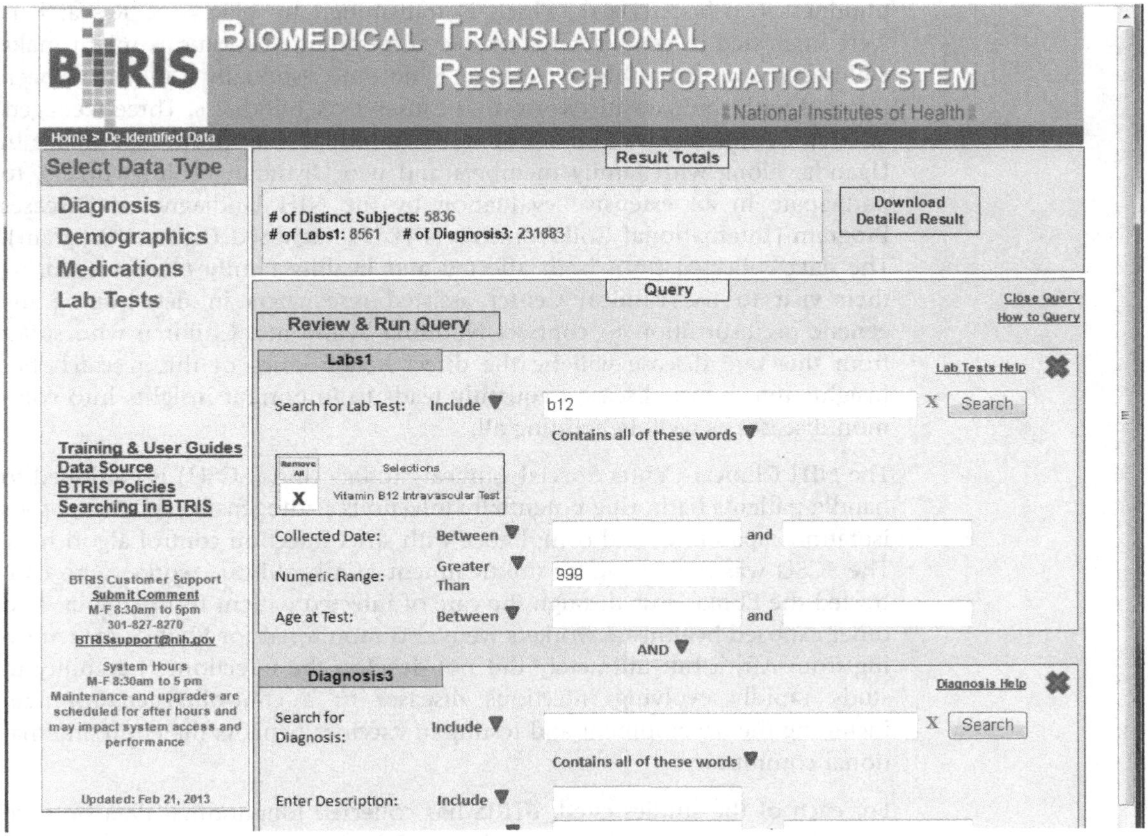

**FIGURE 5.5**
BTRIS deidentified query.

data on 539,000 subjects. Over 11,000 of these subjects are from outside of the United States representing countries from across the globe.

There are many examples of international subjects participating in clinical trials at the NIH Clinical Center. One such study from the 1990s involved family members from a village in Italy who carried the apo A-IMil (Milano) mutant form of apolipoprotein A-I (Roma et al., 1993). Despite hypoalphali-poproteinemia and low levels of HDL, these subjects did not exhibit an increase in cardiovascular diseases. This early research into a small cohort of international subjects led to a new understanding of apolipoprotein A-I function and metabolism.

More recently, the CDC, working with the Ugandan Ministry of Health, studied the potential relationship between Nodding Syndrome and river blindness (onchocerciasis), which is transmitted by blackflies. Researchers were interested in genetic variations or autoimmune conditions might make children more susceptible to Nodding Syndrome, especially if they have been infected with the parasitic worm that causes river blindness. Three teenagers affected by this disease came to the NIH Clinical Center from their village in Uganda, along with family members and two Ugandan health workers, to participate in an extensive evaluation by the NIH Undiagnosed Diseases Program (International Collaborative; NIH Undiagnosed Diseases Program). The data collected from both affected and healthy family members during their visit to the Clinical Center assisted researchers in determining the genetic predisposition to contract Nodding Syndrome. Children who suffer from this rare disease will be the direct beneficiaries of this research but insights into a rare disease frequently leads to important insights into common diseases as well, benefitting all.

The NIH Clinical Center Special Clinical Studies Unit (SCSU) is equipped to handle patients harboring potentially infectious pathogens based on rigorous isolation capabilities and compliance with strict infection control algorithms. The SCSU was the setting for the treatment of a healthcare worker who contracted the Ebola virus through the care of infected patient from Liberia. Two other exposed healthcare workers were also monitored for Ebola after returning from Africa but ultimately did not develop the infection. The ability to study rapidly evolving infectious diseases in a controlled environment including the development and testing of vaccines benefits the entire international community.

For each of the studies cited, BTRIS has collected longitudinal data on each of the participating subjects from both NIH Clinical Center systems as well as institute research systems. Research teams associated with active protocols are able to access data, text, and images within hours for protocol management and analysis. Data are available to researchers for hypothesis testing after identifiers have been removed allowing access to extensive longitudinal datasets. The use of natural language processing and text deidentification software allows for utilization of key data elements from clinical documentation.

## GENERAL REMARKS

A discussion of global collboration on human research participants data must start from an understanding of the relevant national and international laws, policies and procedures regarding the ethical conduct of research and clinical care in the research setting, including, among others, informed consent, patient privacy, protection of research data including electronic data privacy, and research oversight. An interoperable platform that supports storage, analysis, and controlled sharing of genomic and clinical data will in each jurisdiction need to adhere to applicable regulations and standards. To the extent that platforms support data used for patient care, they will also be subject to the laws governing the provision of clinical services.

Legal frameworks and regulatory requirements differ substantially both within and across national borders. Privacy protections on personal data vary considerably. Some policies relate specifically to health or genetic information, whereas others apply broadly to personal information, inclusive of health information. Such legal differences require an organized effort to enable international collaborations to share data across borders. Thus, to the extent that the international scientific and medical communities believe that responsible sharing of data will be key to future progress, there is a pressing need to engage the public and relevant governmental authorities on a collaborative effort to harmonize policies, procedures, and regulations across jurisdictions. This will require a firm foundation of trust from stakeholders regarding protection of privacy, autonomy and subject rights, respect for local jurisdiction, and coordination of global research on ethical, legal, and socialImplication issues, and commitment to the mission of improving human health and patient care. This delicate and multifaceted endeavor must ultimately be guided by the basic principles of a civil society, and respect for diversity of opinions and values among individuals and countries.

To meet the needs of global patient, research, and clinical communities, it will also be necessary to create the technology platforms with open standards creating the technology foundation for:

1. scalable upload and storage of data;
2. rapid processing with state-of-the-art generic and custom tools;
3. management of security, privacy and user access; and
4. downloading and controlled sharing of data and results.

The platforms should have standards and application programming interfaces to securely interact with the data and results, enabling a wide variety of operating entities to serve users. Similar to the World Wide Web, the platform standards must be globally distributed, ubiquitously available, precisely defined, and sufficiently reliable so that both for-profit and not-for-profit organizations will be enabled to securely build upon it.

To realize the benefits of data sharing and integration for the improvement in medical knowledge and human health, it is perhaps best to create a global alliance inclusive of varied stakeholders to develop shared ethical and regulatory policies and procedures, and to create and then to manage interoperable technology platforms with open standards. The alliance should strive to serve the needs of society, rather than solely commercial or academic interests; support current and future technologies for generation of data and for distributed computing; encourage innovation and diversity by engaging both nonprofit and for profit stakeholders; and remain an open resource to the research and clinical communities, rather than serving the needs of any particular entity.

## References

Cimino, J.J., 2000. From data to knowledge through concept-oriented terminologies: experience with the Medical Entities Dictionary. J. Am. Med. Inform. Assoc. 7 (3), 288–297.

Cimino, J.J., 2012. The false security of blind dates: chrononymization's lack of impact on data privacy of laboratory data. Appl. Clin. Inform. 3 (4), 392–403.

Cimino, J.J., Ayres, E.J., Remennik, L., Rath, S., Freedman, R., Beri, A., et al., 2014. The National Institutes of Health's Biomedical Translational Research Information System (BTRIS): design, contents, functionality and experience to date. J. Biomed. Inform. 52, 11–27.

Cimino, J.J., Remennick, L., 2014. Adapting a Clinical Data Repository to ICD-10-CM through the use of a Terminology Repository. AMIA Annu. Symp. Proc. 2014, 405–413.

Dokholyan, R.S., Muhlbaier, L.H., Falletta, J.M., Jacobs, J.P., Shahian, D., Haan, C.K., et al., 2009. Regulatory and ethical considerations for linking clinical and administrative databases. Am. Heart. J. 157 (6), 971–982.

International Collaborative Research Effort Brings Ugandan Family to the Clinical Center. NIH Clinical Center Profile 2015, p. 20. <http://clinicalcenter.nih.gov/about/profile/pdf/Profile_2015.pdf> (accessed 28.08.16).

Nadkarni, P.M., 1997. QAV: querying entity-attribute-value metadata in a biomedical database. Comput. Methods Progr. Biomed. 53 (2), 93–103.

NIH Undiagnosed Diseases Program. <https://www.genome.gov/27544402> (accessed 28.08.16).

Niland, J.C., Rouse, L., Stahl, D.C., 2006. An informatics blueprint for healthcare quality information systems. J. Am. Med. Inform. Assoc. 13 (4), 402–417.

Roma, P., Gregg, R., Meng, M., Ronan, R., Zech, L., Franceschini, G., et al., 1993. In vivo metabolism of a mutant form of apolipoprotein A-I, apo A-IMil,8nf, associated with familial hypoalphalipoproteinemia. J. Clin. Invest. 91 (4), 1445–1452, <http://dx.doi.org/10.1172/JCI116349. http://www.jci.org/articles/view/116349>.

Weber, G.M., Murphy, S.N., McMurry, A.J., Macfadden, D., Nigrin, D.J., Churchill, S., et al., 2009. The Shared Health Research Information Network (SHRINE): a prototype federated query tool for clinical data repositories. J. Am. Med. Inform. Assoc. 16 (5), 624–630.

# CHAPTER 6

# Clinical Information Systems

**P. Degoulet[1,2], D. Luna[3] and F.G.B. de Quiros[3]**

[1]Paris Descartes University, Paris, France
[2]INSERM UMR 1138-E22: Information Sciences & Personalized Medicine, Paris, France
[3]Hospital Italiano de Buenos Aires, Buenos Aires, Argentina

## INTRODUCTION

Information and communication technologies have evolved significantly, actively influencing our society. In the healthcare domain several driving forces have affected the development of health information technology (HIT) applications. They include the evolution of the technology itself and of the underlying healthcare system, the increased financial pressure on care providers, and the progressive citizen involvement in the emerging of a global eSociety (Van de Velde, 1992; Haux, 2006; Vogel, 2014; Ozbolt et al., 2014; Degoulet, 2014; Blum, 1986; Gardner, 2016).

Within hospitals the earliest hospital information systems (HIS) were developed in the mid-1960s in the United States and a few European countries. Their development has followed the general evolution of computer technology (Fig. 6.1): main frames, minicomputers tied together into distributed systems, Internet-based applications, and more recently cloud computing and connected objects (Van de Velde and Degoulet, 2003). The architecture of the earliest mainframe-based systems allowed the connection of health professionals to databases of shared patient-care data. Beginning in 1964 with Lockheed Martin, El Camino Hospital was the first hospital to deploy a computerized provider entry (CPOE) system on top of a shared medical record. The emergence of the mini-computer industry was at the origin of multiple applications dedicated to structures smaller than an entire hospital, in particular in ancillary departments such as biology, radiology, pharmacy, and in some specialized medical units (e.g., intensive care, cardiology, or oncology). More user-friendly interfaces provided by personal computers (PCs) facilitated the access not only to HIS functions but also to personal applications

**129**

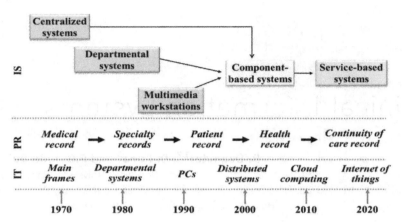

**FIGURE 6.1**

Evolution of health/hospital information systems. *Updated from Degoulet, P., 2014. Hospital information systems. In: Venot, A., Burgun, A., Quantin, C. (Eds.). Medical Informatics, e-Health. Springer-Verlag, Paris, pp. 289–313.*

such as text processing or spreadsheets. PCs became multimedia workstations hiding the complexities of the underlying infrastructure (Degoulet, 2014). In the years 2000 information systems could be progressively built up as sets of software components exchanging standardized messages through exchange platforms (Sauquet et al., 1994; Vogel, 2014). With the development of the Internet and high-speed wide area networks, database servers could quit the hospital environment to be hosted in public or private secured clouds. Patient records that were centralized in mainframe applications, distributed in departmental systems could be unified again within HIS and eventually be consolidated at multisite, regional or even wider areas to constitute integrated delivery networks (IDNs) (Vogel, 2014; Ozbolt et al., 2014). Development of successive generation of digital assistants (personal digital assistants, hand-held PC, smart-phones, and tablets) facilitate the ubiquitous access to health records. Explosive development in the years 2010s of connected objects allow the integration through telecommunication of citizen generated data into the EHR systems and contribute to the emerging big data and Internet of things era.

A clinical information system (CIS) is traditionally defined as the subset of the entire HIS directly devoted to patient care (Blum, 1986; Van de Velde and Degoulet, 2003). Within the hospital boundaries CISs concern both inpatient and outpatient care (Table 6.1). They aim to guarantee that health-related information is provided to the right person at the right time and place to optimize the healthcare quality, safety, accessibility, equity, and efficiency. Initially targeted to support health professionals in their daily

**Table 6.1** A Typology of Healthcare Information Systems (HCIS)

| Information systems | Health Care Organization (HCO) | | | |
|---|---|---|---|---|
| | Physician office/ Practice group | Single institution (Hospital, Rehabilitation center, . . .) | Home care | Integrated care delivery networks (HMO, MCO, ACO) |
| Electronic health record system | AMRS | EHR | EHR | EHR |
| Clinical information system | | CIS | CIS | CCMS |
| Hospital information system | | HIS | HCMS | |
| Integrated delivery system/Health information system | | | | IDS |

*ACO, accountable care organization; AMRS, ambulatory medical record system; CCMS, continuity of care management system; CIS, clinical information system; EHR, electronic health record; HCMS, home care management system; HMO, health maintenance organization, MCO, managed care organization.*

activities, they become progressively patient-centered care systems to take into account the features of empowered patients (Ozbolt et al., 2014). Maturity models such as HIMSS analytics EMRAM (Electronic Medical Record Adoption Model) are becoming standards for HIS/CIS deployment strategies and benchmarking among institutions that have embraced HIT solutions (HIMSS Analytics, 2015). In the United States , e.g., the meaningful use (MU) legislation that provides government subsidies when the adequate use of IT solutions is demonstrated has had a tremendous effect on the rapid diffusion of certified HIS/CIS solutions (Blumenthal and Tavenner, 2010; ONC, 2013; Jones et al., 2014). For example, in the last quarter of 2015, 31.3% of the 5454 US hospitals studied by HIMSS had reached Stage 6 or 7 of the EMRAM maturity model (HIMSS Analytics, 2015). Similar trends, although delayed in time, are observed for certified ambulatory medical record systems (AMRSs).

Industrial offers for hospitals have become mature and progressively concentrate around a limited number of significant offers (Koppel and Lehmann, 2015) at the same time than their nature change with the evolution of the underlying national healthcare systems (Sittig and Singh, 2010). Hospitals are prone to merge to improve their attractiveness and financial efficiency. They establish contracts with private practice consortium offering access to their outpatient management functions that progressively replace existing AMRSs (Table 6.1). Similar strategy frequently applies for home care management that can be driven by a hospital consortium. IDNs combine under the same umbrella various combinations of providers such as private practice groups, acute and postacute care hospitals, or home care facilities. They take the form in the United States of health maintenance organizations (HMO), managed care organization (MCO), or accountable care organizations (ACO) with their underlying state or federal regulation.

This chapter is dedicated to CISs, mainly from a hospital point of view. Two use cases are described, the HIBA (Hospital Italiano de Buenos Aires) and the HEGP (Hôpital Européen Georges Pompidou) in Paris that share the same EMRAM 6 EHR maturity level but differ by their internal organization, their relationships to the broader health information systems, and the underlying national healthcare system.

## CISs FUNCTIONS AND SUBSYSTEMS

Four main groups of functions are necessary to be available in a CIS: patient identification and admission/discharge/transfer functions (ID-ADT), electronic health record (EHR) management, act management (AM), and appointment and resource scheduling (ARS). They can be provided as subsystems of an enterprise resource planning platform or as separate components integrated through a middleware platform (Van de Velde and Degoulet, 2003).

Patient identity management is essential to guarantee the uniqueness of the patient record within one or several institutions and that each professional is able to access a single record whether the patient is examined as inpatient, outpatient, or given advice at distance (i.e., in the form of virtual consultations). Computerized tools allow the search for close identities to avoid the constitution of duplicates. Temporary identities need to be created (e.g., urgency situation) for not yet identified patients but also programs to merge pieces of records belonging to a unique patient. The admission-discharge-transfer (ADT) part is necessary to instantly localize patients within one or several institutions and achieve real-time bed management.

The unique and shared multimedia EHR is the core of any CIS (McDonald et al., 2014) so that the EHR subsystem/component is often used as synonym of CIS. The EHR allows the recording of patient-related data and information whatever their source (physician, nurse, medical device, etc.) or their nature (e.g., clinical, biological, imaging, nursing, images, "omics" data, procedures, treatments). Access to patient information should be accessed at any time, any place by whom it concerns, and only whom it concerns.

The AM subsystem, more frequently denominated as CPOE subsystem/component, allows the healthcare professionals, according to their user profile (e.g., physician vs nurse, senior vs junior), to prescribe individual acts (e.g., serum potassium, vital sign follow up), groups of acts constituting an order set (e.g., ionogram), or more complex associations that will be specific of a clinical situation (e.g., urinary infection, coronary dilation) or a disease stage (e.g., chemotherapy for a given type of cancer). Branching rules (e.g., starting and ending condition for a drug) and/or loops can be described in the most complex protocols (e.g., a chemotherapy or a radiotherapy protocol). Orders are stored in the EHR and allow the production of the different care plans for patients or groups of patients (e.g., set of beds in a clinical unit). Strong integration of the EHR and CPOE is needed to generate the various reminders/alerts/alarms/suggestions needed for the optimal/personalized management of patients. Close-loop drug delivery is achieved when the complete circuit is secured starting from the prescription and ending with the drug dispensing. Bar-coding or radiofrequency identification (RFID) techniques help checking that the right dose is given at the right time to the right patient by the right person (e.g., nurse) holding an appropriate identification system (e.g. bracelet for the patient, badge for the nurse, bar-code for the drug container).

The ARS subsystem/component allows the efficient management of a single or a set of appointments for a patient (e.g. radiotherapy or dialysis sessions) or a group of patients. Managing the different appointments (inpatient/outpatient visits, specialist advices, investigations, operating rooms; virtual visits) by the same tool within an organization facilitates the optimal use of resources and the integration of the patient preferences and geographical constraints (e.g., time to go from point A to point B within a hospital or between two healthcare facilities) when planning the appointments (e.g., a 1-day medical or surgical stay). Close integration among the EHR, CPOE, and ARS components are necessary to manage complex protocols integrating multiple appointments and branching conditions. Capacity for professionals external to institution and/or the patient to obtain appointments through the Internet is a growing demand as well as providing general information about the hospital and its services.

Behind these four major subsystems/business components, additional subsystems/components are necessary to allow the smooth functioning of a CIS. The authorization component includes the description of the various CIS end-users including their profile and rights. Data protection and data/functions access procedures (i.e., access rights) need to be defined according to the profession of the user (e.g., physician vs nurse vs secretary, senior vs junior), the point of access (inside/outside the hospital), the role of the professional as regard the care of the patient, and the presence/absence of the patient in the institution at the time of access. The traceability of accesses should be guaranteed with a granularity sufficient in case of controversies between the patient and the institution. CISs have to be adapted to the regional/national regulations that are likely to change over time and in particular the security/confidentiality rules.

The reference manager is a generic component that needs to be fed with health-related specific nomenclatures/terminologies, as well as data and knowledge structures (e.g., data entry forms, exchanged message structures, hospital drug/devices repositories, structured protocol description).

For example, diagnosis codes are likely to rely on the 9th or 10th revision of the international classification of diseases (ICD9/ICD10). LOINC is the current standard for biological orders terms (LOINC, 2016). SNOMED CT is the most comprehensive nomenclature of medical terms (SNOMED-CT, 2016). HL7 and more recently FHIR are international messaging standards (e.g., admission/discharge/transfer procedures, exchange messages between the EHR and the laboratories or the pharmacy) (HL7 v2/v3, 2016; Benson and Grieve, 2016). DICOM is the imaging standard in medicine including structures for their acquisition conditions, the storage of images, and the associated reports (DICOM, 2016). The HL7 v3 Clinical Document Architecture (CDA) is a document markup standard that specifies the structure and semantics of "clinical documents" to be exchanged between healthcare providers and patients (HL7 CDA, 2016).

The HL7 CCOW (Clinical Context Object Work-group) standard keeps among other functions track of the patient and user identification on a given terminal (the context of work), to help the possibility of making decision errors by displaying the records of several patient at the same time (CCOW, 2011).

The decision support system (DSS) component can be triggered by most major components and particularly the EPR and CPOE components. Domain-independent business rules syntaxes such as SBVR (Semantics of Business Vocabulary and Business Rules) can be used in addition to more domain-dependent business rule syntaxes such as Arden or GLIF (Hripcsak et al., 1994; Ohno-Machado et al., 1998; Boussadi et al., 2011).

Integration within the components of a CIS can be achieved by access to shared database structures (e.g. the patient record structures) or by messaging. Communication tools are necessary in case of message communication. IHE (Integrating the Healthcare Enterprise) promotes the coordinated use of established standards through integration profiles (IHE, 2016). When different terminologies coexist in a CIS, the reference manager may help in solving semantic interoperability issues by mapping the terms of these terminologies (i.e., acting as a terminology server). Integration of ancillary applications needs the development of various plug-in on top of the applications and semantic interoperability tools such as semantic mediators (Wiederhold, 1992; Degoulet et al., 1998).

## ORGANIZATIONAL ASPECTS

IT governance or eGovernance represents the adaptation to IT of general governance principles. COSO, CobiT, and ITIL are significant examples of reference frameworks for IT governance (COSO, 2016; COBIT, 2016; ITIL, 2016). Several dimensions are included in these frameworks. They include IT strategic alignment, resource management, risk management, performance and services management, communication and change management, and various evaluation aspects such audit and control, use and satisfaction, maturity, IT value and return on investment (ROI) measurements.

IT strategic alignment is the first step of a value-generating process at the enterprise level. For a hospital it might mean better quality of care, better attractiveness, and greater productivity (Haux et al., 2004). The IT strategic plan that derives from this task frequently concern a 3- to 5-years period for which the IT strategic plan should fully comply with the global institution strategic plan (Bopp and Brown, 2005). Examples of targets at the hospital level include the development of a quality strategy, the reduction of the total number of beds, the development of alternatives to traditional inpatient care (e.g., one-day hospital care, outpatient activity, home care), the acquisition of complementary structures and/or the hospital integration into a larger consortium. How starting from an initial IT state A to reach a final state B will support the institution strategic plan is the key objective of any IT strategic plan. Both institution and IT strategic plans need to be validated by the institution governance committee (Fig. 6.2).

CISs depend on interrelated sets of elements consisting of peoples, activities, and material resources interacting to process data, information, and knowledge. To achieve organizational objectives, CIS aim to distribute and exchange information in the most efficient manner, providing communication between the various stakeholders and the patients. HIS/CIS project management require a high level of leadership to facilitate policymaking and ensure the project's governance. Within a hospital/institution, IT management should preferably be controlled by an eGovernance committee attached to the hospital/institution governance or strategic committee. The role of this committee is to guarantee the alignment of the IT strategy with the overall hospital strategy and to report to the general governance committee on the state of evolution of the CIS. The IT commission is the place where the IT manager(s) should discuss with all categories of end-users involved with the CIS. Ad hoc subgroups can be created for any given specific project.

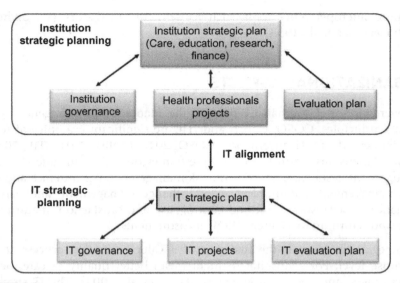

**FIGURE 6.2**
IT alignment within healthcare organizations.

(e.g., change management, telemedicine applications, patient empowerment, big data, and data analytics projects).

Resource organization within the IT department depends on the CIS maturity level achieved, but also of the strategic choice of using inhouse versus off the shelf components, as well as internal versus external public/private data storage capacities. Overall IT management is traditionally attributed to a chief information officer (CIO) who belongs to the eGovernance committee and if possible to the institution strategic committee. The need for a CMIO (Chief Medical Information/Informatics Officer) was raised in the late 1980s as an essential support for the development and deployment of CISs (Friedman and Martin, 1988; Haux et al., 2004; Kannry et al., 2016). The CMIO is most frequently but not systematically a physician with certified training in clinical informatics. The $29 billion HITECH incentive and MU legislation have dramatically accelerated the engagement of CMIOs in the United States with employment rates raising from 30% in 2006 to 76% in 2012 (Kannry et al., 2016). The CMIO acts as a project leader/coordinator for the CIS deployment strategy and may report to the hospital chief executive officer (CEO), to the chief medical officer (CMO) or to the CIO. He is the key intermediate between the IT department, the health professionals inside the institution, and the various external CIS vendor(s).

Chief Nursing Information/Informatics Officers positions are more and more frequently proposed in hospital with mature CISs. Nurses represent the most prevalent health profession and interfere in most care processes (Collins et al., 2015). CNIOs can refer to the chief nursing officer (CNO), to the CIO, or to the CMIO. The CNIO is active in the IT committee may be a permanent member of the eGovernance committee. Chief Pharmacy Information/Informatics Officers (CPIO) and Chief Research Informatics Officers (CRIO) have been appointed in some hospitals. To embrace the leading role of health professionals with certified IT roles, a dedicated AMIA task force proposed to group them under the general cover of Chief Clinical Informatics Officer (CCIO) to lead all related CISs' activities (Kannry et al., 2016).

## CIS MONITORING AND EVALUATION

Continuous monitoring/evaluation of CIS project advancement, use, and end-user satisfaction is an essential part of any information system project (Ammenwerth et al., 2003). It needs to be performed at each phase of a CIS project, i.e., before installation, during the deployment phase, and later at the consolidation and MU stages when all users are supposed to adequately use all CIS components and major usability flaws have been corrected (Hadji et al., 2016; Hadji and Degoulet, 2016). It should be integrated into a more global evaluation strategy including outcome measures as well as financial ROI (Meyer and Degoulet, 2008). Acceptance models concentrate on the determinants of user satisfaction and CIS use, whereas success models integrate outcome indicators of CIS deployment such indicators of quality of care or financial ROI (Fig. 6.3).

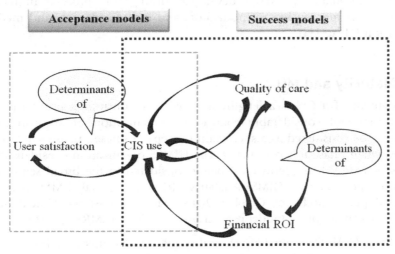

**FIGURE 6.3**
Acceptance and success models.

Evaluations should rely on validated models based on sets of criteria grouped around several evaluation dimensions. Examples of frequently used models include the Information System Success Model (ISSM) of DeLone and McLean (DeLone and McLean, 2003; Petter and McLean, 2009) the Davis Technology Acceptance Model (TAM) with its extensions (TAM2, UTAUT) (Davis, 1989; Lee et al., 2003; Venkatesh et al., 2003), and the Bhattacherjee Expectation Confirmation Model (ECM) of information system continuance (Bhattacherjee, 2001). Data recording can be achieved through group interviews, questionnaires, direct measurements, ethnographic studies including video recording, or a combination of these methods.

The selected dimensions and their importance depend on the deployment stage of the information system to be evaluated (Hadji et al., 2016). In the earliest CIS deployment phases, it is important to observe professional users as close as possible to their working environment in order to evaluate their ease of using the CIS and the interaction process (i.e., system usability), as well as the CIS performance, the quality of the initial training, the quality of support (e.g., materials' repair or bug corrections), and to a more general extent how the CIS addresses end-users' expectations. In a CIS MU phase, satisfaction might depend on the personal characteristics of end-users, the quality of the system, and its flexibility to adapt to the changes required by well-trained end-users (Hadji and Degoulet, 2016).

Fig. 6.4 illustrates the dimensions retained to evaluate the HEGP CIS since the opening of the hospital in July 2000 (Hadji and Degoulet, 2016) and the associated research hypotheses around the unified model of information system continuance (UMISC) developed during this process. Individual characteristics are considered moderators and include age, sex, and medical profession ($H_a$ to $H_d$).

## CIS Maturity and MU

Deployment of a CIS is a lengthy and resource consuming process (3−10 years) that will considerably change the organization of an institution. It should be considered as a succession of steps or phases, phase $n + 1$ being started when phase $n$ is completed. The HIMSS organization considers seven stages of EMRAM, the term EMR being considered to the broad sense used in this chapter as a CIS (HIMSS Analytics, 2015). The global HIMMS Analytics EMR adoption model of October 2015 that combines the United States and European adoption model is described as follows (EMRAM, 2015):

- Stage 0: All three ancillary clinical systems are not installed (i.e., laboratory, radiology, pharmacy) not installed
- Stage 1: All three ancillary clinical systems installed

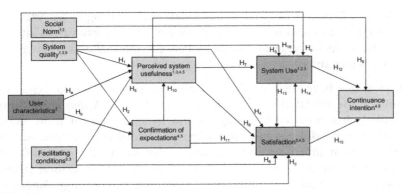

**FIGURE 6.4**

The UMISC developed at HEGP (adapted from Hadji, B., Degoulet, P., 2016. Information sytem end-user satisfaction and continuance intention: a unified modeling approach. J. Biomed. Inform. 61, 185–193). Evaluation dimensions are selected from previously validated models. 1 = Technology Acceptance Model 2 (TAM2) (Davis, 1989; Lee et al., 2003), 2 = Unified Theory of Acceptance and Use of Technology (UTAUT) (Venkatesh et al., 2003), 3 = Information System Success Model (ISSM) (DeLone and McLean, 2003; Petter and McLean, 2009), 4 = Expectation Confirmation Model (ECM) (Bhattacherjee, 2001), 5 = Information Technology Post Adoption Model (ITPAM) (Palm et al., 2010). *HEGP*, Hôpital Européen Georges Pompidou; *UMISC*, Unified Model of Information System Continuance.

- Stage 2: Major ancillary clinical systems feed a clinical data repository (CDR) that provides physicians access for reviewing all orders and results. The CDR contains a controlled medical vocabulary. Rudimentary conflict checking is provided by the clinical decision support/rule engine (CDS). Information from the imaging system may be linked to the CDR at this stage.
- Stage 3: Nursing/clinical documentation (e.g. vital signs, flow sheets, nursing notes,) is integrated with the CDR for at least on inpatient service in the hospital. Implementation of the first level of clinical decision support (CDSS) (e.g. drug/drug, drug/lab conflicts). The electronic medication record functions are implemented. The PACS is available outside radiology.
- Stage 4: CPOE for use by authorized professionals. Second level of CDSS related to evidence-based medicine protocols.
- Stage 5: Fully integrated PACS including the cardiology PACS available through the institution intranet.
- Stage 6: Full physician documentation with structured templates and discrete data for at least on inpatient service in the hospital. Third level of CDS with variance and compliance alerts. Close-loop medication administration with bar coded or RFID to guarantee the five rights of drug administration (i.e., right patient, right drug, right dose, right route, right time).

- Stage 7: Paperless hospital. Clinical data warehousing to analyze pattern of care and improve the quality of care, patient safety, and care delivery efficiency. Data continuity between internal and ambulatory care is guaranteed.

EMRAM stages have their origin in the initial periods of CIS deployment focusing on the availability of functions (Stages 0 to 3). A significant number of hospitals has now reached Stage 5 and over seeking for more detailed description of possible steps to reach level 7. Availability of most functions in one inpatient unit does not guarantee extension to the overall hospital which implies that domain-specific issues should be solved. Critical examples include urgency departments, intensive care units, or oncology departments where complex protocols and clinical rules need to be tested and implemented.

## THE HOSPITAL ITALIANO EXPERIENCE IN BUENOS AIRES

### Argentina's Healthcare System

The health system of Argentina is the result of successive reforms which added new models, without currently achieving a single system and ensuring universal health coverage. The public system which provides free coverage is only used by low-income peoples or during accidents and disasters to provide early care. Health authority and policymaking, in the public system, is delegated to the provinces (federal system). There is also a Social Security system for formal workers, similar to the German system. Private insurance system close to the HMO system in the United States is available for people that are willing to pay for dual coverage.

This way, Argentina has different sectors of insurance and service provision that are approximately distributed as follows: 40% from the public system, 50% from the social security, and 10% from private prepaid care, with a large overlap between these sectors in both the financing and delivery of services. Health expenditure represented 9.4% of the GDP in 2001 and only 4.8% in 2014.

### The Hospital Italiano Care Network

HIBA is a private nonprofit organization, founded in 1853, which currently has become a healthcare network which includes 2 hospitals, 41 operating rooms, 800 home care beds, 25 primary and outpatient care clinics, 250 private offices, and a dental health network. This healthcare network is distributed in the city of Buenos Aires and its metropolitan area. It provides health services for the social security system and an extensive list of private health insurers and cooperatives. Through its "Plan de Salud" (PS), HIBA operates as an insurer, collecting voluntary payments of about 160,000 members. The total network has about 750 beds and employs around 8400 people. The main HIBA hospital employs around 7100 peoples, of whom about 1700 are physicians and 1400 are nurses (Table 6.2).

## Strategic Planning

The ambitious HIBA strategic plan that began in 1998 included the redesign of its service delivery network, a strategy to develop skills in pursuit of excellence, continuity and coordination of care policy, and significant investment in infrastructure and advanced medical equipment. The associated IT strategic plan first stage (1998−2003) began with the inhouse development and implementation of a shared EHR in the outpatient setting (Fig. 6.5). The second stage of this IT plan (2004−09) was focused on the development of the common inpatient and outpatient record, the adoption of standards for health information interoperability, and development of terminology services. Patients were given access to their record through a personal health record. The current phase (3rd) of the IT plan (2010−) is characterized by the improvement of human−computer interface, the complete implementation of the CPOE with CDSS (in every hospital setting), and the initiation of telemedicine and mobile health projects.

**Table 6.2** Italiano and HEGP Hospitals, Main Figures

| Hospital Characteristics | Hospital Italiano, Buenos Aires (Main Hospital) | HEGP |
|---|---|---|
| Total number of active beds (inpatient + 1-day care) | 600 | 810 |
| Nb. intensive care units beds | 160 | 90 |
| Number of operating rooms | 34 | 24 |
| Total number of employees (FTE) | 7100 | 3260 |
| Number of senior physicians (FTE) | 1700 | 490 |
| Number of nurses (FTE) | 1400 | 1100 |
| Nb. of inpatient admissions | 36,000 | 92,900 |
| One-day hospital care (% of admissions) | 58 | 64 |
| Nb. of inpatient admissions with surgical intervention(s) | 28,000 | 24,400 |
| Number of patient visits | 2,400,000 | 200,000 |
| Nb. visits at the emergency department/day | 900 | 140 |
| Nb. PC/laptops/light terminals | 3700 | 3200 |

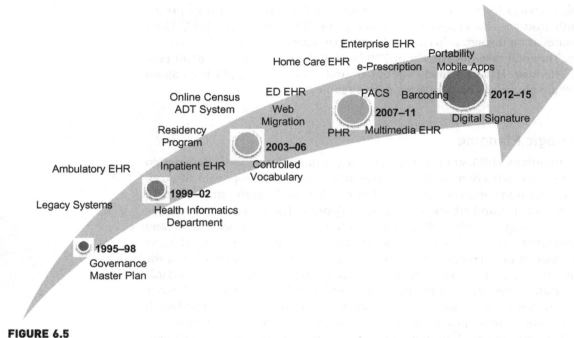

**FIGURE 6.5**

HIBA CIS development. *HIBA,* Hospital Italiano de Buenos Aires; *CIS,* clinical information system.

## The HIBA CIS

The main component of the HIBA CIS (ITALICA) is the common EHR that interoperates with administrative and ancillary applications, through an interoperability bus. The EHR follows a problem-oriented organization that covers outpatient and inpatient care, emergencies, and home care (Aguilera Díaz et al., 2010). Clinical information concerns patient history, preventive information, orders, investigation results, procedures report (including surgeries and anesthetics), and medication lists. Order and order sets that are effected through the CPOE and close-loop delivery are in the design phase (Taliercio et al., 2014). Appointment scheduling is achieved at the integrated network level.

Integration and interoperability issues were considered essential from the beginning of the CIS project and a dedicated working group initiated in 2000. HL7 and DICOM standards were used for results and image transmission and access all over the institution (Soriano et al., 2010). The terminological control strategy included the use of SNOMED CT as reference terminology and the creation of an interface vocabulary which allowed automatic coding for problems, diagnoses, procedures, and drugs (Osornio et al., 2007). Terminology management functions were grouped into a terminology service that was progressively made available on a transnational basis (Gambarte et al., 2007; Luna et al., 2010). Using the terminology services, problem lists and discharge summaries are automatically coded in SNOMED CT. A useful complement was the addition of MedlinePlus patient-related knowledge using the info buttons standard in the personal health record (Strasberg et al., 2013; Borbolla et al., 2014). HL7 CDA was used to standardize the production of reports, starting with laboratory and radiology studies, and currently every health encounter. Reports are stored in a document base (SNOMED coded CDA documents), which is kept in the central EHR repository (Campos et al., 2013).

### Organization

The CIS project is driven by a physician specialized in information systems, with professionals from different areas and disciplines that make up a cross-disciplinary working team. CIS-associated projects are managed with the Department of Health Informatics. The concurrent medical residency program that was initiated in 2001 was found invaluable to build up the necessary professional expertise.

### Evaluation

HIBA CIS was certified HIMSS EMRAM level 6 in April 2016. This achievement was preceded by the constitution of an accreditation team consisting in two physicians and one nurse from the clinical informatics group, working in tandem with the development group and an experienced EMRAM accreditation consultant, who were in charge of planning and carrying out the CIS adaptations necessary for the HIMSS accreditation. Consistency within terminology and use of structured templates for data entry were considered as strengths of the system.

Evaluation studies concerned the qualitative evaluation of nurse expectations as regard the future deployment of the close-loop drug delivery system (Taliercio et al., 2014), and the search for predictive variables of the use of personal health records by HIBA affiliated patients.

Efficiency of the terminology services in coding every input string encouraged us to offer this service to other institutions, not only in our country but in healthcare networks from Chile and Uruguay (Luna et al., 2010).

## THE HEGP EXPERIENCE IN PARIS

### The French Healthcare System

The French healthcare system has two main characteristics, first a statutory universal coverage that concerns more than 99% of a population of 66.6 million habitants in 2016, and second a free choice for patients of health providers (Chevreul et al., 2015; OECD, 2015). Health expenditure represented 9.9% of the GDP in 2001 and 11.5% in 2014, in the highest values of European countries but remained much lower than the United States corresponding figures of 13.7% and 17.1% (The World Bank, 2016). Sixty-one percent of hospital are public and represent 78% of beds. Hospital budget is mainly on diagnostic-related group (DRG) coding with adjustment for teaching and research activities. Ambulatory physicians' income is based on a fee for service with for subgroups pay for performance or capitation adjustments. Seventy-nine percent of expenses are taken in charge by the universal covering systems, 13% by complementary insurances, and 7% correspond to patient out of pocket payment.

### The HEGP Hospital

HEGP is an 800-bed acute care public university hospital located in southwest Paris that opened in July 2000 after the merging of three aging facilities, the Boucicaut, Broussais, and Laennec acute and postacute care hospitals. It is one of the 38 hospitals managed by the AP-HP public organization, with a total number of 95,000 employees. HEGP is organized around three major healthcare cooperating centers: cardiovascular, cancer, internal medicine including an emergency department and trauma center. A total of 3260 employees, including the equivalent of 1100 registered nurses and 490 full-time-equivalent physicians staff the hospital (Table 6.2). Observed activity in 2015 is summarized in Table 6.2. A total of 92,900 admissions were registered of which 61% were for one-day inpatient care (medical or surgical). On a daily basis, about 140 patients are routinely examined at the emergency and trauma center. HEGP is associated with two postacute and long-term care hospital (Corentin-Celton and Vaugirard) to constitute the HUPO group (Hôpitaux Universitaires Paris Ouest). Twelve hospital groups like the HUPO group administration structure, AP-HP (Assistance Publique Hôpitaux de Paris) are in charge of the management of the 38 structures.

### Strategic Planning

Strategic planning for the project took place during 1995 and 1996, and the project was started in December 1996. A best-of-breed approach was adopted, with a focus on integration and communication between predefined business components. Preliminary extracts from the HISA model

(European prestandard PrEnv 12967–1) were included in the technical documentation (CEN TC251, 1997).

After an 18-month selection process, a consortium was selected, headed by SYSECA, a branch of the French Alcatel company (later belonging to THALES), with Hewlett Packard as the major hardware vendor and MEDASYS and Per-Sé Technologies (now MainCare) as software component providers (Degoulet et al., 2003). A 5-year contract was signed on April 1998, which included a $24 \times 7$ guarantee that integrated solutions would be operational at the opening of the hospital and thereafter. The role of the prime contractor was to provide project support for the duration of the contract; to provide the necessary healthcare-related components, either directly or through subcontracts; and to provide the middleware tools for the integration of the different components into a three-layered architecture. The contract was renewed every 4 years since 2003.

## The HEGP CIS

Since its opening in July 2000, HEGP has therefore been equipped with a fully integrated CIS. This CIS consists of five major components from three different providers integrated by the middleware platform (Degoulet et al., 2003). Three components are based on the DxCare software suite from Medasys: (1) a Patient identification and ADT component, (2) an electronic multimedia shared health record (EHR), and (3) a CPOE component, including single orders and order sets (e.g., biology, radiology, drug prescriptions, or nursing care prescriptions) as well as more complex ones (e.g., protocol-based) covering all possible categories of acts. One component based on OneCall from McKesson/Maincare solutions is used for all categories of resource and appointment scheduling (e.g., in/outpatient visits, operating rooms, radiotherapy sessions). The fifth component, i.e.., the clinical data warehouse (CDW), is based on the Informatics for Integrating Biology and the Bedside (i2b2) that integrates all patient-oriented structured and non-structured data from the opening of the HEGP (Zapletal et al., 2010; Murphy et al., 2010). With the exception of the CDW, which was operational in 2010, the four other components have been operational since the opening of the hospital. Computerized orders from the CPOE and results from ancillary applications (biology, pharmacy, radiology, and pathology) are transmitted through standardized messages. The middleware layer includes a shared dictionary of concepts to foster semantic integration between the different healthcare components, the communication manager, the decision support engine, the document manager, and an HL7 clinical context of work (CCOW) manager. Major improvements since 2000 have concern usability issues, and the improvement of the EHR integrated DSS (e.g., alarms, complex protocol description, and management).

### Organizational Aspects

Hospital strategy is defined on a 5-year basis in the strategic planning business report discussed and validated by the hospital general governance committee. It includes an IT strategic plan so that IT strategy can be aligned. The IT strategic plan need to comply with the general AP-HP strategic plan. The CMIO, a physician with IT training, was also appointed as CIO. The CIO/CMIO was a permanent member of the hospital governance committee. The IT governance committee includes the hospital CEO, CMO, CIO/CMIO, and CRIO.

### Evaluation

HEGP CIS was certified HIMSS/EMRAM level 6 in May 2013. Six successive electronic evaluation satisfaction surveys were performed in 2004, 2008, 2011, and every year since 2013 (Hadji et al., 2016) using subsets of the UMISC model described in (Hadji and Degoulet 2016). These surveys were based on self-administered questionnaires consisting of 51–58 questions. Self-reported CIS use was assessed for each of the 12–18 functions considered. One to seven Likert scales (1 = not available, not used or not appropriate, 2 = very rarely, 3 = rarely, 4 = occasionally, 5 = rather frequently, 6 = frequently, 7 = very frequently) were used for all dimension-related evaluation questions. A significant increase in CIS use was observed between 2004 and 2015, while overall satisfaction increased regularly, even in the very late post adoption period, as showed in Fig. 6.6.

In multiple regression analysis and structural equation models (Hadji et al., 2016), bidirectional relationships between satisfaction and use were significant in the early deployment phase (4 years) but disappeared in the late (8 years) and very late phases (>10 years). In a nearly paperless environment,

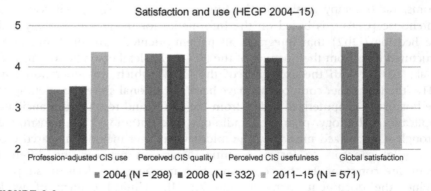

**FIGURE 6.6**

Evaluation of CIS main acceptance dimensions during three successive phases at HEGP. Self-reported use and acceptance dimensions are evaluated on 1–7 scales. *CIS*, clinical information system; *HEGP*, Hôpital Européen Georges Pompidou.

all users access the CIS whether satisfied or not suggesting that disappearance of the bilateral relationship between CIS use and end-user satisfaction could be considered as an indirect sign of CIS maturity.

The HEGP CDW directly fed from the CIS (Zapletal et al., 2010) has been extensively used for clinical and translational research (e.g., patient selection in the context of a more personalized practice of medicine) (Canuel et al., 2015; Rance et al., 2016). In the pharmaceutical domain at the HEGP the EHR/CDW integration was found invaluable in the evaluation of the frequency of both drug prescription and drug dispensing errors (Caruba et al., 2010; Berdot et al., 2012), the impact of alerts on drug dosage adjustment (Sellier et al., 2009), the role of pharmacy drug order validation (Estellat et al., 2007), but also to automatically detect drug–drug interactions (Girardeau et al., 2015) or to perform in silico evaluations of decision rules for further integration into the operational environment of the CPOE (Boussadi et al., 2012).

## DISCUSSION AND CONCLUSION

Development and deployment of CISs are long-term initiatives that can span over several decades. The two use cases described in this chapter have reached a similar degree of functionality despite disparities in their organization and their underlying healthcare system. In term of functionalities both CISs rely on a shared multimedia record covering both inpatient and outpatient care and made of both free text and structured information derived from questionnaires adapted to the different specialties of the hospital. Share terminology services foster semantic integrity not only within the EHR but also between the EHR and other component that need to access the EHR such as the DSS. Terminology services allow the integration of standard nomenclatures such as ICD10, SNOMED, or LOINC. National-based terminologies include DRG categories used for financing in France and drug nomenclatures in both countries. Home care management is integrated within the HIBA system but managed independently for HEGP patients. Exchanges between components rely on standardized messages mainly HL7 based.

Early integration of the IT strategic plan into the global institution plan was considered essential in both cases. This means among others considering IT development as a major hospital objective, involving top level management in the different phases of the CIS project (i.e., as main actors in the IT governance committee), and integrating the CIO as a permanent member of the hospital governance committee. Health professionals with dual training were hired early within the IT department including the equivalent of the currently named CMIO. Involvement of nurses within was considered essential in both cases with a CNIO position defined from the beginning at the HEGP.

Resistance to change was higher at HEGP for the medical personnel than for the nursing one, in particular when the CPOE functions were introduced. Usability studies such as the one performed at the HEGP generated multiple requests for user interface improvements that were later implemented and validated. Automatic coding from free text, such as the one developed at HIBA might reduce health professional workload. In a highly research-oriented organization as HEGP early development of CIS-associated CDW was considered as a highly supporting strategy from a physician point of view.

Perspective common to the two institution include better integration with the external health professional world but clearly with the patient with the development of the many functions required such as the easy access to the institution resources (e.g., appointment scheduling, patient record access), the development of telemedicine applications (e.g., home-based), the integration of patient generated data (e.g., pre-visit formularies, connected objects), and the provision of more and more personalized care.

## References

Aguilera Díaz, J., Arias, A.E., Budalich, C.M., et al., 2010. Development and implementation of an integrated EHR for Homecare Service: a South American experience. Stud. Health Technol. Inform. 160 (Pt 1), 43−47.

Ammenwerth, E., Mansmann, U., Iller, C., Eichstadter, R., 2003. Factors affecting and affected by user acceptance of computer-based nursing documentation: results of a two-year study. J. Am. Med. Inform. Assoc. 10, 69−84.

Benson, T., Grieve, G., 2016. Principles of health interoperability, SNOMED CT, HL7 and FHIR. third revised ed. Springer, New York.

Berdot, S., Sabatier, B., Gillaizeau, F., Caruba, T., Durieux, P., 2012. Evaluation of drug administration errors in a teaching hospital. BMC Health Serv. Res. 12, 60, <http://dx.doi.org/10.1186/1472-6963-12-60>.

Bhattacherjee, A., 2001. Understanding information systems continuance: an expectation-confirma- tion model. MIS Q. 25 (3), 351−370.

Blum, B., 1986. Clinical Information Systems. Springer Verlag, Berlin, vol. 1, 413 pages.

Blumenthal, D., Tavenner, M., 2010. The "meaningful use" regulation for electronic health records. N. Engl. J. Med. 363 (6), 501−504.

Bopp, K.D., Brown, G.D., 2005. Aligning information strategy and business and clinical strategies: information as a strategic asset. In: Brown, G.D., Stone, T.T., Patrick, T.B. (Eds.), Strategic Management of Information Systems in Healthcare. Health Administration Press, Chicago, IL, pp. 121−147.

Borbolla, D., Del Fiol, G., Taliercio, V., Otero, C., Campos, F., Martinez, M., et al., 2014. Integrating person- alized health information from MedlinePlus in a patient portal. Stud. Health Technol. Inform. 205, 348−352.

Boussadi, A., Caruba, T., Zapletal, E., Sabatier, B., Durieux, P., Degoulet, P., 2011. A business rules design framework for a pharmaceutical validation and alert system. Methods Inf. Med. 50 (1), 36−50.

Boussadi, A., Caruba, T., Zapletal, E., Sabatier, B., Durieux, P., Degoulet, P., 2012. A clinical data ware-house-based process for refining medication orders alerts. J. Am. Med. Inform. Assoc. 19 (5), 782–785.

Campos, F., Plazzotta, F., Luna, D., Baum, A., FGB, De. Quirós, 2013. Developing and implementing an interoperable document-based electronic health record. Stud. Health Technol. Inform. 192, 1169.

Canuel, V., Rance, B., Avillach, P., Degoulet, P., Burgun, A., 2015. Translational research platforms integrating clinical and omics data: a review of publicly available solutions. Brief Bioinform. 16 (2), 280–290.

Caruba, T., Colombet, I., Gillaizeau, F., et al., 2010. Chronology of prescribing error during the hospital stay and prediction of pharmacist's alerts overriding: a prospective analysis. BMC Health Serv. Res. 10, 13, <http://dx.doi.org/10.1186/1472-6963-10-13>.

CCOW, 2011. HL7-CCOW standard description. <http://www.hl7.org/implement/standards/product_brief.cfm?product_id=1> (accessed 31.05.16).

CEN TC251, March 1997. Healthcare Information System Architecture Part 1 (HISA). Healthcare Middleware Layer. prENV 12967-1. Brussels: CEN TC251. <http://www.centc251.org>.

Chevreul, K., Brigham, K.B., Durand-Zaleski, I., Hernandez-Quevedo, C., 2015. France: health system review. Health Syst. Trans. (European Observatory) 17 (No.3).

COBIT. Cobit 5 framework. <http://www.isaca.org/COBIT/Pages/default.aspx> (accessed 31.05.16).

Collins, S.A., Alexander, D., Moss, J., 2015. Nursing domain of CI governance: recommendations for health IT adoption and optimization. J. Am. Med. Inform. Assoc. 22, 697–706.

COSO. COSO official website. <http://www.coso.org/> (accessed 31.05.16).

Davis, F., 1989. Perceived usefulness, perceived ease of use, and user acceptance of information technology. MIS Q. 13, 319–340.

Degoulet, P., 2014. Hospital information systems. In: Venot, A., Burgun, A., Quantin, C. (Eds.), Medical Informatics, e-Health. Springer-Verlag, Paris, pp. 289–313.

Degoulet, P., Sauquet, D., Jaulent, M.C., Zapletal, E., Lavril, M., 1998. Rationale and design considerations for a semantic mediator in health information systems. Methods Inf. Med. 37 (4-5), 518–526.

Degoulet, P., Marin, L., Lavril, M., Le Bozec, C., Delbecke, E., Meaux, J.J., et al., 2003. The HEGP component-based clinical information system. Int. J. Med. Inform. 69, 115–126.

DeLone, W.H., McLean, E.R., 2003. The DeLone and McLean model of information systems success: a ten-year update. J. Manage. Inform. Syst. 19c (4), 9–30.

DICOM, 2016. DICOM NEMA main website. <http://medical.nema.org/> (accessed 31.05.16).

EMRAM, 2015. Global EMRAM model. <http://www.himssanalytics.org/research/emram-stage-criteria> (accessed 06.06.16).

Estellat, C., Colombet, I., Vautier, S., Huault-Quentel, Durieux, P., Sabatier, B., 2007. Impact of pharmacy validation in a computerized physician order entry context. Int. J. Qual. Health Care 9 (5), 317–325.

Friedman, B.A., Martin, J.B., 1988. The physician as a locus of authority, responsibility, and operational control of medical systems. J. Med. Syst. 12 (6), 389–396.

Gambarte, M.L., Osornio, A.L., Martinez, M., Reynoso, G., Luna, D., de Quiros, F.G.B., 2007. A practical approach to advanced terminology services in health information systems. Stud. Health Technol. Inform. 129 (1), 621–625.

Gardner, R., 2016. Clinical information systems—from yesterday to tomorrow. Yearb. Med. Inform. S62–S75, <http://dx.doi.org/10.15265/IYS-2016-s010> Special 25th Anniversary Edition.

Girardeau, Y., Trivin, C., Durieux, P., Le Beller, C., Louet Agnes, L.L., Neuraz, A., et al., 2015. Detection of drug-drug interactions inducing acute kidney injury by electronic health records mining. Drug Saf. 38, 799–809, <http://dx.doi.org/10.1007/s40264-015-0311-y>.

Hadji, B., Degoulet, P., 2016. Information sytem end-user satisfaction and continuance intention: a unified modeling approach. J. Biomed. Inform. 61, 185–193.

Hadji, B., Martin, G., Dupuis, I., Campoy, E., Degoulet, P., 2016. 14 Years longitudinal evaluation of clinical information systems acceptance. Int. J. Med. Inform. 86, 20–29.

Haux, R., 2006. Health information systems—past, present, future. Int. J. Med. Inform. 5 (3–4), 268–281.

Haux, R., Winter, A., Ammenwerth, E., Brigl, B., 2004. Strategic Information Management in Hospitals. Springer-Verlag, New York, NY.

HIMSS Analytics, 2015. <http://www.himssanalytics.org/home/index.aspx>.

HL7 CDA, 2016. HL7-CDA® R2 standard description. <http://www.hl7.org/implement/standards/product_brief.cfm?product_id=7> (accessed 31.05.16).

HL7 v2/v3, 2016. HL7 Version 2/3 message standard description. <http://www.hl7.org/> (accessed 31.05.16).

Hripcsak, G., Ludemann, P., Pryor, T.A., Wigertz, O.B., Clayton, P.D., 1994. Rationale for the Arden syntax. Comput. Biomed. Res. 27 (4), 291–324.

IHE, 2016. Main IHE web page. <http://www.ihe.net/> (accessed 31.05.16).

ITIL. ITIL official website. <https://www.axelos.com/best-practice-solutions/itil> (accessed 31.05.16).

Jones, S.S., Rudin, R.S., Perry, T., Shekelle, P.G., 2014. Health information technology: an updated systematic review with a focus on meaningful use. Ann. Intern. Med. 160 (1), 48–54.

Kannry, J., Sengstack, P., Thyvalikatath, T.P., et al., 2016. The Chief Clinical Informatics Officer (CCIO). AMIA Task Force Report on CCIO Knowledge, Educations, and Skillset Requirements. Appl. Clin. Inform. 7, 143–176.

Koppel, R., Lehmann, C.U., 2015. Implications of an emerging EHR monoculture for hospitals and healthcare systems. J. Am. Med. Inform. Assoc. 22 (2), 465–471, <http://dx.doi.org/10.1136/amiajnl-2014-003023>.

Lee, Y.A., Kozar, K.A., Larsen, K.R.T., 2003. The technology acceptance model: past, present and future. Commun. Assoc. Inf. Syst. 12 (50), 752–780.

LOINC. [http://loinc.org/] (LOINC site, accessed May 31, 2016).

Luna, D., Lopez, G., Otero, C., Mauro, A., Casanelli, C.T., de Quirós, F.G.B., 2010. Implementation of interinstitutional and transnational remote terminology services. AMIA Annu. Symp. Proc. 482–486.

McDonald, C.J., Tang, P.C., Hripcsak, G., 2014. Electronic health record system. In: Shortliffe, E.H., Cimino, J.J. (Eds.), Biomedical Informatics. Springer, London, pp. 391–421.

Meyer, R., Degoulet, P., 2008. Assessing the capital efficiency of healthcare information technologies investments: an econometric perspective. Yearb. Med. Inform. 114–127.

Murphy, S.N., Weber, G., Mendis, M., Chueh, H.C., Churchill, S., Glaser, J.P., et al., 2010. Serving the enterprise and beyond with informatics for integrating biology and the bedside (i2b2). J. Am. Med. Inform. Assoc. 17 (2), 124–130.

OECD, 2015. Health at a Glance 2015: OECD Indicators. OECD Publishing, Paris, <http://dx.doi.org/10.1787/health_glance-2015-en>.

Ohno-Machado, L., Gennari, J.H., Murphy, S., Jain, N.L., Tu, S.W., Oliver, D.E., et al., 1998. The GuideLine Interchange format: a model for repreenting guidelines. J. Am. Med. Inform. Assoc. 5 (4), 357–372.

ONC, 2013. <http://www.healthit.gov/providers-professionals/how-attain-meaningful-use>.

Osornio, A.L., Luna, D., Gambarte, M.L., Gomez, A., Reynoso, G., de Quirós, F.G.B., 2007. Creation of a local interface terminology to SNOMED CT. Stud. Health Technol. Inform. 129, 765–769.

Ozbolt, J., Bakken, S., Dykes, P.C., 2014. Patient-centered care systems. In: Shortliffe, E.H., Cimino, J.J. (Eds.), Biomedical Informatics. Springer, London, pp. 475–501.

Palm, J.M., Dart, T., Dupuis, I., Leneveut, L., Degoulet, P., 2010. Clinical information system post-adoption evaluation at the Georges Pompidou university hospital. AMIA Annu. Symp. Proc. 582–586.

Petter, S., McLean, E.R., 2009. A meta-analytic assessment of the DeLone and McLean IS success model: An examination of IS success at the individual level. Inform. Manage. 46 (3), 159–166.

Rance, B., Canuel, V., Countouris, H., Laurent-Puig, P., Burgun, A., 2016. Integrating heterogeneous biomedical data for cancer research: the CARPEM infrastructure. Appl. Clin. Inform. 7 (2), 260–274.

Sauquet, D., Jean, F.C., Lemaitre, D., et al., 1994. The HELIOS Unification Bus: a toolbox to develop client/server applications. Comput. Methods Programs Biomed. 45 (Suppl), S13–S22.

Sellier, E., Colombet, I., Sabatier, B., et al., 2009. Effect of alerts for drug dosage adjustments in inpatients with renal insufficiency. JAMA 16, 203–210.

Sittig, D.F., Singh, H., 2010. A new sociotechnical model for studying health information technology in complex adaptive healthcare systems. Qual. Saf. Health Care 19 (Suppl. 3), i68–i74.

SNOMED-CT. SNOMED-CT documentation site. <http://www.ihtsdo.org/fileadmin/user_upload/doc/> (accessed 31.05.16).

Soriano, E., Plazzotta, F., Campos, F., et al., 2010. Integration of healthcare information: from enterprise PACS to patient centered multimedia health record. Stud. Health Technol. Inform. 160 (Pt 1), 126–130.

Strasberg, H.R., Del Fiol, G., Cimino, J.J., 2013. Terminology challenges implementing the HL7 context-aware knowledge retrieval ('Infobutton') standard. J. Am. Med. Inform. Assoc. 20 (2), 218–223.

Taliercio, V., Schachner, B., Borbolla, D., Luna, D., Villalba, E., Quiros, F., 2014. The expectations of nurses about the implementation of a Barcoded Medication Administration System: a qualitative study. Stud. Health Technol. Inform. 205, 191–195.

Van de Velde, R., 1992. Hospital Information Systems—The Next Generation. Springer, New York, NY; Berlin, vol. 1, 472 pages.

Van de Velde, R., Degoulet, P., 2003. Clinical Information Systems—A Component-Based Approach. Springer, New York, NY, vol. 1, 294 pages.

Venkatesh, V., Morris, M., Davis, G., et al., 2003. User acceptance of information technology: toward a unified view. Manag. Inf. Syst. Q. 27, 425–478.

Vogel, L.H., 2014. Management of information in health care organizations. In: Shortliffe, E.H., Cimino, J.J. (Eds.), Biomedical Informatics. Springer, London, pp. 443–474.

Wiederhold, G., 1992. Mediators in the architecture of future information systems. Computer 25 (3), 38–49.

The World Bank. Health expenditure, total (% of GDP). <http://data.worldbank.org/indicator/SH.XPD.TOTL.ZS> (accessed 28.04.16).

Zapletal, E., Rodon, E., Grabar, N., Degoulet, P., 2010. Methodology of integration of a clinical data warehouse with a clinical information system. Stud. Health Technol. Inform. 160 (Pt 1), 193–197.

# CHAPTER 7

# The New Medicine: From the Paper Medical Record to the Digitized Human Being

**D. Sigulem[1], M.P. Ramos[1] and R. de Holanda Albuquerque[2]**
[1]Federal University of São Paulo — UNIFESP, São Paulo, Brazil
[2]Brazilian Diabetes Association, Brazil

## CONTENTS

## THE DIGITIZATION OF HEALTHCARE INFORMATION

In 1990 the first computers arrived in the medical environments—offices, clinics, and hospitals—creating great expectations among its users. What could be done with them? From the existing basic software at that time—a word processor, spreadsheets and a database—it was possible to develop simple applications that would, for instance, transpose the patient's medical records partially to an electronic media.

Despite the general familiarity with the medical records on paper and the little training necessary to deal with it—since it was used for a long time by every healthcare professional—despite its low cost and reasonably easy mobility within the institutions, its disadvantages were obvious: many notes were illegible, they could only be in one place at a time, they could be sometimes incomplete or ambiguous and were not able to generate alerts. The recorded data was not structured or standardized and any research aiming a clinical protocol, from one patient or a group of them, was very difficult to be carried through.

Therefore the development of the electronic medical record (EMR), also known as electronic health record (EHR), the latter more comprehensive, has always been the dream of medicine and of health professionals everywhere in the world. The EHR has emerged as the repository of information of the health status of an individual and the care it was provided to him/her throughout life, electronically stored in a way that could serve multiple and legitimate users of the system (Mc Donald et al., 2014). This multimedia electronic record allows visualization and analysis of data, images, and signals; it can generate different types of alerts and enables quick definition of trends.

The health informatics community envisioned the design and development of the single electronic patient record, available to all authorized health professionals in all health institutions, i.e., an universal record. However, due to problems related to the lack of consistency of medical language, the existence of different operational systems, the strong intervention of the health industry—protecting their own patient record systems, both in the clinic and hospital settings —and the lack of communication between the private and public systems, the dream of an universal electronic patient record was practically buried.

In March 2001 the Institute of Medicine published a report called *Crossing the Quality Chasm: A New Health System for the 21st Century*, pointing out the state of health information technology: "[...] health care organizations, hospitals, and physician groups typically operate as separate 'silos' acting without the benefit of complete information about the patient's condition, medical history, services provided in other settings, or medications provided by other clinicians." (IOM, 2001, p. 2).

The aimed solution is the interoperability between existing systems. Some pilot projects were launched in 2011 in the United States, involving the largest companies that develop electronic records for healthcare, in an attempt to adopt open models (open-source models) that could break through the barriers of interoperability (Lohr, 2011).

According to Topol (2012), an alternative to electronic records in healthcare is the personal health record (PHR) system, where the patient's data is recorded in a flash drive or in a cloud or both. "The advantage is that the patient becomes responsible for collecting and storing every bit of data produced by every doctor, by every exam taken, and by every healthcare system used."

PHR is often seen as a means of communication between patients and healthcare providers and as a way to engage patients in understanding their health and subjects related to it. However, to meet the demands of the patient-centered care, there must be space for changes in the processes and standards of clinical practice, in the organization and management of health services, and in the education of patients and health professionals (Ozbolt et al., 2014).

The three main PHRs developed in the United States so far were: Microsoft's Health Vault,[1] Google Health,[2] and WebMD Health Manager.[3] In addition to these, a portal called MiCare[4] was developed to be a gateway between the patient's data archive of his choice and other features of the PHR. The portal has been well-accepted, but the researchers behind this project cited four important lessons learned:

1. Even though patients could request the transfer of their data from EHR to their PHR, the speed of the transfer was very slow.

2. Healthcare professionals insisted on a minimum period of 7 days to register information in the EHR, time needed to get in contact with the patient and explain the results of an exam, for instance.
3. Partial information in the system could jeopardize a correct clinical decision.
4. In the same way, to give to the patient the power of deciding which data should be shared with the providers could also threaten the correct interpretation of a case by a professional, left to take decisions without vital information (Mc Donald et al., 2014).

Nevertheless the true power of the technology of healthcare information goes beyond health records (Topol, 2012). The report *Realizing the Full Potential of Health Information Technology to Improve Healthcare for Americans: The Path Forward*, prepared by the President's Council of Advisors on Science and Technology (PCAST) in December 2010, analyzes "[...] how health information technology (HIT) could improve the quality of health care and reduce its cost, and whether existing Federal efforts in health information technology are optimized for these goals" (PCAST, 2010, p. 9):

> Health information technology can allow clinicians to have real-time access to complete patient data, and provide them with support to make the best possible decisions. It can help patients become more involved in their own care, which is especially important in managing chronic conditions like diabetes, asthma, or heart disease. It can enable a range of population level monitoring and real-time research such as the detection of developing epidemics, health risks in the environment, or adverse events caused by medications. It can improve clinical trials, leading to more rapid advances in personalized medicine. It can streamline processes and reduce administrative overhead, as it has in other industries. It can lead to the creation of new, high-tech markets and jobs. Finally, it can help support a range of economic reforms in the healthcare system that will be needed to address our country's long-term fiscal challenges. As David Blumenthal, the National Coordinator for Health Information Technology, has written, "Information is the life blood of modern medicine, [and] health information technology is destined to be its circulatory system.

The moment we are living is the consequence of a specific, contemporary way of engaging with healthcare, which is diffused and involves more people than ever before. The patient's house and the community where he/she lives (family, work colleagues, friends, neighborhood etc.) are fast becoming the places where healthcare is provided.

The necessary technologies of information to account for the needs of patients and their caregivers should not only migrate from hospitals and other institutions to the community, but they should also be fed with information resources to aid patients in complex decision-making processes and in their communication with health professionals. Access to all this data should bring patients more familiar with clinical records and improve knowledge of the science of health needed to help them understand their own conditions and participate in appropriate treatment (Consumer Health Informatics and the Personal Health Records; Johnson et al., 2014).

## DIGITAL MEDICINE

According to Elenko (2015), the term Digital Medicine was first coined in 2002 by Shaffer (2002), and can be defined as: "[...] that technology and those products that are undergoing rigorous clinical validation and/or that ultimately will have a direct impact on diagnosing, preventing, monitoring or treating a disease, condition or syndrome."

Several areas in medicine can be directly impacted with the inclusion of technological innovations (Elenko et al., 2015):

- *Continuous and remote monitoring*: Pandemics controlling and diseases surveillance; blood pressure monitoring, healthcare applications on mobile devices; physical exercise follow-up.
- *The digital phenotype*: Study of biological markers that may help to prevent diseases such as Alzheimer, or to monitor their progress; signal processing and machine learning are being used to predict apnea in children within intensive care units; in the context of the consumer, the devices are registering, among other signs, caloric intake and sleep patterns.
- *Remote disease management*: Monitoring and management of chronic diseases, especially cardiovascular, pulmonary, neurological and geriatric; monitoring of diabetes and obesity, through custom digital programs;
- *The connected patient*: In addition to the data collected via personal devices, patient communities are proactively providing information about their own experiences and treatments. Examples: PatientsLikeMe,[5] Alliance Health's,[6] CureTogether.com[7] of genetic testing company 23andMe.[8]
- *Security and privacy*: One of the greatest challenges of digital medicine to regulate when and how collected and stored data should be shared or reused.

## mHEALTH AND THE e-PATIENT

Tom Ferguson coined the term e-patients to describe individuals who are **equipped, enabled, empowered and engaged** in their health and healthcare decisions. He envisioned healthcare as an equal partnership between e-patients and health professional and systems that support them.

**Ferguson (2007).**

The ubiquity of mobile phones, cloud computing, social media, and open data sources are putting healthcare in the hands of every citizen and by doing so, setting it free from medical mainframes (Fig. 7.1).

Besides, mobile phones were adopted by consumers all over the world much faster than previous innovations. The PC took 16 years to be adopted by one-quarter of the US population. Mobile phones took 13 years and the web took only 7 years (Fig. 7.2).

The infiltration of mobile phones in our daily lives and their technological evolution, coupled to data communication services and internet connections that are getting faster and broader every day, are present throughout the world. In February 2016 the Brazil's National Telecommunications Agency (Anatel) announced that the country closed the year 2015 with 257.8 million mobile lines (Anatel, 2015).

This presence is changing the way people communicate. There is a clear preference nowadays for text messaging instead of vocal conversations. Hence the success of networks like Facebook©, Twitter©, and WhatsApp©. The migration of the use of voice to data reduces the cost of interoperability. Less costs for the user, more profit for the operators.

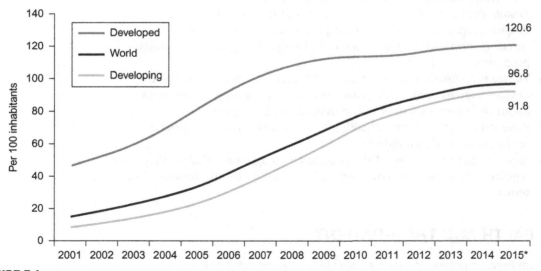

**FIGURE 7.1**

Mobile cellular subscriptions per 100 inhabitants from 2001 to 2015 in the United States. *http://www.itu.int/en/ITU-D/Statistics/Pages/stat/default.aspx (accessed 04.05.16).*

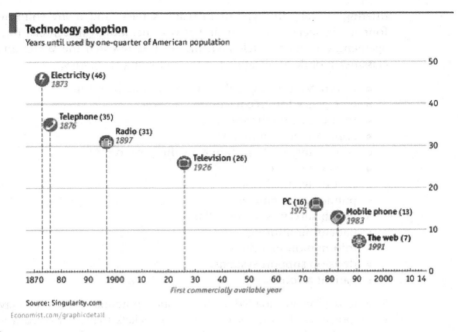

**FIGURE 7.2**

Technology adoption. Year until used by one-quarter of American population. *The Economist. http://www.economist.com/blogs/ graphicdetail/2014/03/daily-chart-7?%3Ffsrc%3Dscn%2F = tw%2Fdc. Accessed 05/04/2016.*

Furthermore, people are using more and more their devices and social networks to seek information about their own health. The healthcare industry believes so strongly in this trend that has already applied millions of dollars in the development of devices (coupled or not to smartphones), to collect information about vital signs from patients and in the development of applications for smartphones (apps) that analyze these vital signs (IMS, 2013).

According to the *World Health Organization's Second Global on Health Survey* (WHO, 2011), mHealth is defined as:

> mHealth is a component of eHealth. [...] mHealth or mobile health as medical and public health practice supported by mobile devices, such as mobile phones, patient monitoring devices, personal digital assistants (PDAs), and other wireless devices.
>
> mHealth involves the use and capitalization on a mobile phone's core utility of voice and short messaging service (SMS) as well as more complex functionalities and applications including general packet radio service (GPRS), third and fourth generation mobile telecommunications (3G and 4G systems), global positioning system (GPS), and Bluetooth technology.

The research (WHO, 2011) says that "there is a grounds well of activity" in the use of mobile health technologies, with 83% of WHO's Member States offering at least one type of mHealth service, and many countries offering four to six specific mHealth initiatives, including call centers, disaster management, and mobile telemedicine. In this research the WHO analyzes and classifies mobile applications in the following categories:

- health call centers/health care telephone help line;
- emergency toll-free telephone services;
- treatment compliance;
- appointment reminders;
- community mobilization & health promotion;
- raising awareness;
- mobile telemedicine;
- public health emergencies;
- health survey sand surveillance;
- patient monitoring;
- information initiatives;
- decision support systems;
- patient records

Mobile devices are loaded with so much potential that, some say, they are more powerful than the state-of-the-art rockets that discovered a new planet recently. People today carry in their pockets more information and data than any library in the world could possibly store. And there is already a concept that defines this behavior: to load information on your device is called BYOD + K, i.e., BringYourOwnDevice.[9] The K means knowledge.

As a result, the e-patients today reach the environments of their consultations with a lot of information, some of which are not even known by those responsible for their care (Shortliffe and Blois, 2014). The way a consultation takes place has also changed. Not always a person who is typing in the cell is inattentive to what the doctor says: he/she might be just checking online what has been said to prepare better for a participation in the conversation.

The mobile technology revolution, allied to Internet access and social networking, has impacted greatly medical practice and the way healthcare is being pursued. The notion of relationship is being fundamentally redefined and this alters greatly the role of all parts involved in healthcare: doctors, patients, companies, public and private financers, etc.

There are currently many low-cost technologies available to a great deal of the population—e-patients and e-doctors (TMF, 2016):

*Low-cost ubiquitous sensors*:

1. *The smartphone*: We can perform various diagnostic measures such as ECG, heart rate, respiratory rate, blood pressure, oxygen saturation, recognition of the skin lesions, retinal examination, etc.

2. *Biochemical test sensors*: The best known and simplest to use is the glucose meter; the more complex so far are the ones that recognizes dengue fever disease through a PCR chip (polymerase chain reaction) (Tan et al., 2014).
3. Low cost versions of devices that were once costly, such as portable ultrasound, which can be used on a large scale and for a variety of clinical situations.

*Health Communities*: where patients exchange information and data with each other. Those spaces placed in the hands of people a much larger amount of information than some big computers could ever carry. Wireless communications and satellites allowed the dissemination of information, knowledge, and the formation of social networks in so-called computer clouds.

But there is still a lot of uncertainty among physicians regarding the use of applications on a daily basis. One thing certain is that there is still a long way to integrate them in clinical practice. Aitken (2013) analyzed 43,700 existing applications in the US Apple iTunes app store in June 2013 and found that 54% were genuinely medical applications. Another 9% were directed to the general public (consumers) and 31% were for clinical use.

The applications aimed at consumers, in turn, had a simple design and gave only information. One hundred and fifty captured data, but only fifty offered to manage clinical conditions. The author concludes that these applications still have a long road ahead of them to become accepted in professional practice.

A research developed at the Manhattan Medical Research[10] shows the most used applications and websites by doctors within the population researched:

*Mobile apps*: Eighty-one percent of smartphone owners use at least one application; and the 5 most used professional apps were:

- *Epocrates Apps*: Point of Care Medical applications[11] designed to provide information about drugs to doctors and other healthcare professionals;
- Web MD professional App[12]: It provides health information and tools for health managing;
- UpToDate[13]: It is an evidence-based, physician-authored clinical decision support resource that clinicians trust to make the right point-of-care decisions
- Sermo[14]: It is a social network for doctors. It is private and exclusively for doctors.
- QuantiaMD[15]: It is a web and mobile community for physicians.

*Mobile websites*: Sixty-seven percent of smartphone owners use at least one application; and the 5 most used professional websites were:

- The WebMD Professional Network—includes eMedicine, Medscape, or WebMD
- Epocrates

- NIH Websites—includes Medline/NIH, PubMed, National Cancer Institute (NCI), or NIH
- UpToDate
- Wikipedia

Wicks and Chiauzzi (2015) discussed studies that analyze several healthcare apps, and they found many issues to be tackled, such as poor methodological approaches, lack of documentation, calculation of inappropriate medication doses, lack of quality, problems with security and privacy, among others.

For this reason, the Food and Drug Administration (FDA, 2015) developed a document called *Mobile Medical Applications: Guidance for Food and Drug Administration Staff*, informing the regulation rules for the development of healthcare-related devices and mobile apps.

## DATA USAGE AND SECURITY

As the data of individuals is being accumulated, it is possible to predict behavior and to anticipate preventive measures in a given population. With the so-called big data it is possible to build models and foresee further procedures each person shall need. This is the most important step to be taken toward personalized medicine.

This set of data when treated appropriately can be used in the structuring of the evidence-based medicine, supporting the development of policies and guidelines used by healthcare professionals and by government policies of countries or global health organizations.

On the other hand, as the systems become more sophisticated, concerns about the security of data of health professionals, patients and institutions raise, bringing serious implications for the confidentiality and privacy of citizens. This is a constant topic of discussion in the medical field and there is still no consensus on the issue.

As it was said before, FDA has issued the *Mobile Medical Applications Guidance for Industry and Food and Drug Administration Staff* (FDA, 2015) "[...] which explains the agency's oversight of mobile medical apps as devices and the focus is only on the apps that present a greater risk to patients if they don't work as intended and on apps that cause smartphones or other mobile platforms to impact the functionality or performance of traditional medical devices." However the FDA does not regulate EHS and EPS.

## POINT OF CARE

Point of Care is an essential component for this global view of healthcare described here, which incorporates all the innovative technological opportunities, such as mobile biosensors, portable and intelligent computing technologies, information and communication. The US National Institute of Health (NIH, 2010), in its strategic study on the subject said:

> With the development of miniaturized devices and wireless communication, the way in which doctors care for patients will dramatically change and the role patients take in their own healthcare will increase. Healthcare will become more personalized through tailoring of interventions to individual patients.

Thuemmler (2015) from The Cyber Academy of Edinburgh in Scotland presented in 2015 a vision of how Point of Care is migrating from big institutions such as hospitals to the periphery. The Point of Care could be represented by: tele-health, m-health, community-nurse, hospices, day-surgery, day-clinics, physio-therapists, social care, and psychotherapists.

For the developed countries, these new concepts are seen as an opportunity for reducing costs. Meanwhile, in two-thirds of the underdeveloped countries, this new approach to healthcare is being used to look after populations under poor living conditions, monitoring infectious diseases and epidemics like HIV, tuberculosis, malaria, dengue fever, zika fever, etc.

It is worth noting that the implementation of new technologies and new concepts of health self-care imply considerable behavioral changes, which will require major investments in the education of multiprofessional teams, as well as of the population as a whole.

Basically, doctors consider as their Point of Care the work in their own practice. This space may be physically within or outside a hospital or an outpatient clinic. In the course of their work, they use EMRs; sometimes very simple ones like notebooks, but more recently also quite elaborate records, with decision-making systems, guidelines, instant access to Google and electronic applications. For multidisciplinary teams, the best systems allow the use of a single record where they all write their conduct on the cases. With mobile devices, however, this space was expanded and the health professional can be called at any time by his patients through communication tools online. Doctors and patients communicate by email and more often via SMS1, WhatsApp1, Twitter1, Facebook1, and Instagram1.

## THE FUTURE: A NEW MEDICAL CHART

On November 25, 2015, the United States announced the release of $215 million for a project called Precision Medicine Initiative (PMI, 2015). The project aims to gather and analyze medical and lifestyle data of a million people. This is a new approach to medical research, which seeks to customize the understanding of the disease and its treatment through genetic information of environmental and social conditions. The first diseases chosen were: cancer, diabetes, Alzheimer's, cardiovascular disease, obesity and mental disorders.

The main goal is to try to explain why people who have the same disease respond differently to the same treatment, or how come cancer or cardiovascular diseases also affect people who have a healthy lifestyle.

Christensen et al. (2008), in the book *The Innovator's Prescription: A Disruptive Solution for Healthcare*, defined Precision Medicine (PM) as: "Precision Medicine is an approach to discover and develop medicines, vaccines or routes of intervention (behavior, nutrition, etc.) that enable disease prevention and deliver superior therapeutic outcomes for patients, by integrating 'Big Data', clinical, molecular (multi-omics including epigenetics), environmental and behavioral information to understand the biological basis of a disease."

Also Servant and his collaborators (2014) explained: "PM combines the knowledge of the patient's characteristics with traditional medical records and environmental information to optimize health. PM does not only rely on genomic medicine but also integrates any other relevant information such as non-genomic biological data, clinical data, environmental parameters and the patient's lifestyle."

Fig. 7.3 shows the feasibility of monitoring variables-environment, genomes, biochemical and physiological parameters, and social environment involved in the health state of a person.

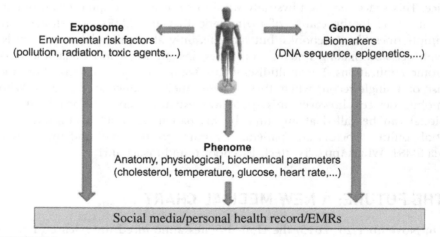

**FIGURE 7.3**

The exposome, the genome, and the phenome contributions to the social medial, PHR, and EMR.
*Modified from Martin-Sanchez, F.J., 2015. The new era of digital medicine: new challenges for Health Informatics. Medical Informatics in Europe. MIE. Madrid <https://www.emaze.com/@AFCWOWCZ/Presentation-Name> (accessed 30.04.16).*

# FINAL CONSIDERATIONS

The digital revolution, which started around 1980 with the Internet and after with mobile devices, social networking, big data, and computing clouds, revolutionized work practices in healthcare.

Patients became more autonomous, confident, and they can do this without having any formal education, training, or support. In the 20th century, many medical educators believed that it was possible to empower the patient giving them information to address certain clinical situations. Today we know that this is not the solution to our health systems.

In other words, we cannot simply replace the old model centered on the figure of the doctor for a solely patient-centered model. Ideally, the near future is looking at a collaborative model where the e-patient of the 21st century will work together with the 21st century doctor using the resources available in the 21st century. It should be emphasized, however, that many features of the doctors of the 20th century—such as personal contact, face-to-face, clinical examination, and emotional assessments should continue to be present in the practice of medicine.

Although medicine is a super conservative area and often resistant to new changes, we can no longer ignore that our lives have been radically transformed by digital innovations. "Radically transformed. Creatively destroyed." (Topol, 2012).

According to Topol (2012), this destruction is now possible in healthcare because, for the first time in history, we are being able to digitize the human being. "Digitizing human beings means knowing in depth their genome, their vital functions, their blood parameters, and everything else that keeps them alive." (Fig. 7.4).

That is, we are living a moment where the development of new *technologies* are converging that power in the ancient medicine, traditional, and sedimented to a new medicine.

Berwick (2016) says "The new directions are centered in the care of people, in the family joint participation, in the doctor's transfer of control, in the compliance with the needs of the poor and in the defense of care as a universal human right."

Ferguson (1995) was already prophesying the changes of medicine from the industrial era to the era of information, pointing out to an increasingly greater autonomy of the patient (Fig. 7.5).

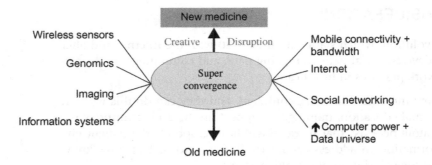

**FIGURE 7.4**

The new medicine after a creative disruption. *Modified from Topol, E., 2012. Electronic health records and health information technology. In: The Creative Destruction of Medicine. How Digital Revolution Will Create Better Health Care. Basic Books, New York, NY. pp 145.*

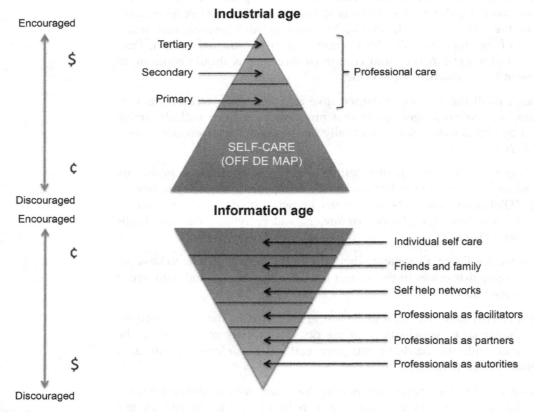

**FIGURE 7.5**

The healthcare in the industrial era and in the information era. *Ferguson, T., 1995. Consumer health informatics. Healthc. Forum J. 38(1), 28–33.*

## BOX 7.1 THE CONVERGENCE OF THE PRECISION MEDICINE AND THE PARTICIPATIVE MEDICINE TO THE PREVENTIVE MEDICINE

| | |
|---|---|
| Precision medicine | • Need of earlier diagnosis |
| | • More personalized therapies |
| | • Clinical trials and the development of new drugs need to be faster and more effective |
| | • Improve disease classification systems |

↓
Preventive
Medicine
↑

| | |
|---|---|
| Participative Medicine (patient centered) | • Risk profiling, disease prediction and prevention |
| | • Control health system costs |
| | • Citizens should take more responsibility for the maintenance of their own health |
| | • Emphasis on prevention, not cure |

*Source: Modified from Fernando J. Martin-Sanchez. <https://www.emaze.com/@AFCWOWCZ/Presentation-Name>.*

In relation to Precision Medicine and Participative Medicine, Martin-Sanchez (2015) pointed out the current challenges of medicine (Box 7.1):

Much remains to be done, from the validation of applications and devices to the verification of the used methodologies of validation, the scientific basis of the process, the safety of individuals and evidence of the cost—benefit of its adoption. But one thing is certain: this is a process with no turning back.

## End Notes

1. https://msdn.microsoft.com/en-us/healthvault/healthvault-introduction.aspx.
2. Discontinued in 2011.
3. http://www.webmd.com/phr.
4. https://eclaims.micaresvc.com.
5. https://www.patientslikeme.com/—Cambridge, MA, USA.
6. https://www.alliancehealth.com/—Salt Lake City, UT, USA.
7. http://curetogether.com.
8. https://www.23andme.com/en-int/—Mountain View, CA, USA.
9. https://en.wikipedia.org/wiki/Bring_your_own_device
10. http://www.manhattanmedicalresearch.com/
11. http://www.epocrates.com/
12. http://www.webmd.com/mobile
13. http://www.uptodate.com/home
14. http://www.sermo.com/
15. https://www.quantiamd.com

# References

Aitken, M., 2013. Patient Apps for Improved Healthcare: From Novelty to Mainstream. Report by the IMS Institute for Healthcare Informatics. <http://obroncology.com/imshealth/content/IIHI%20Apps%20report%20231013F_interactive.pdf> (accessed 30.04.16).

Anatel, 2015. Brasil perde 8% de linhas de celular em 2015, diz Anatel. <http://blogs.estadao.com.br/link/brasil-perde-8-de-linhas-de-celular-em-2015-diz-anatel/> (accessed 30.04.16).

Berwick, D.M., 2016. Era 3 for medicine and health care. JAMA. 315 (13), 1329–1330, <http://dx.doi.org/10.1001/jama.2016.1509>.

Christensen, C.M., et al., 2008. The Innovator's Prescription: A Disruptive Solution for Healthcare. McGraw-Hill, New York, NY, p. 2008.

Elenko, E., Underwood, L., Zohar, D., 2015. Defining digital medicine. Nat. Biotechnol. 33, 456–461.

FDA, 2015. Mobile Medical Applications. Guidance for Industry and Food and Drug Administration Staff.

Ferguson, T., 1995. Consumer health informatics. Healthc. Forum J. 38 (1), 28–33.

Ferguson, T., 2007. e-Patients: how they can help us heal healthcare. <http://www.e-patients.net/e-Patients_White_Paper.pdf> (accessed 30.04.16).

Goodman, K.W., Cushman, R., Miller, R.A., 2014. Ethics in biomedical and health informatics: users, standards, and outcomes. In: Shortliffe, E.H., Cimino, J.J. (Eds.), Biomedical Informatics. Computers Applications in Health Care and Biomedicine, fourth ed. Springer-Verlag, New York, NY, p. 399.

IBGE, 2015. <http://www.ibge.gov.br/apps/populacao/projecao/index.html> (accessed 29.04.16).

IMS, 2013. PatientApps for improved healthcare. From novelty to mainstream. IMS Institute for Healthcare Informatics, October 2013. <http://www.imshealth.com/en/thought-leadership/ims-institute/reports/patient-apps-for-improved-healthcare> (accessed 30.04.16).

IOM. Institute of Medicine, 2001. Crossing the quality chasm: a new health system for the 21st century. <http://nationalacademies.org/HMD/Reports/2001/Crossing-the-Quality-Chasm-A-New-Health-System-for-the-21st-Century.aspx#sthash.WoT6ptMF.dpuf> (accessed 30.04.16).

Johnson, K., Brugge, H.J., Mandl, K.D., 2014. Consumer health informatics and the personal health records. In: fourth ed Shortliffe, E.H., Cimino, J.J. (Eds.), Biomedical Informatics Computers Applications in Health Care and Biomedicine, 2014. Springer-Verlag, New York, NY, p. 523.

Lohr, S., 2011. U.S. tries open-source model for health data systems. The New York Times <http://bits.blogs.nytimes.com/2011/02/02/u-s-tries-open-source-model-for-health-data-systems/?_r = 0> (accessed 30.04.16).

Martin-Sanchez, F.J., 2015. The new era of digital medicine: new challenges for health informatics. medical informatics in Europe. MIE 2015. Madrid <https://www.emaze.com/@AFCWOWCZ/Presentation-Name> (accessed 30.04.16).

Mc Donald, C.J., Tang, P.C., Hripcsack, G., 2014. Electronic health record systems. In: Shortliffe, E.H., Cimino, J.J. (Eds.), Biomedical Informatics Computers Applications in Health Care and Biomedicine, fourth ed. Springer-Verlag, New York, NY, p. 391.

MMR, 2013. Manhathan Medical Researh. Manhattan Research, Taking the Pulse® U.S. 2013, Medscape, Epocrates, Wikipedia, PubMed, UpToDate <https://www.google.com.br/url?sa=t&rct=j&q=&esrc=s&source=web&cd=1&ved=0ahUKEwiCsqLu4rHMAhUouYMKHXr-XD-MQFgggMAA&url=http%3A%2F%2Fwww.ehealthcaresolutions.com%2Fsummit%2Fpresentations%2Fmressi_2013.pptx&usg = AFQjCNEUfxiJ8gSjKN4bP0ua_nBZJXbvNw>.

Ozbolt, J., Bakken, S., Dykes, P.C., 2014. Patient-centered care systems. In: Shortliffe, E.H., Cimino, J.J. (Eds.), Biomedical Informatics: Computers Applications in Health Care and Biomedicine, 4fourh ed. Springer-Verlag, New York, NY, pp. 497–500.

PCAST, 2010. Realizing the full potential of health information technology to improve healthcare for Americans: the path forward. <https://www.whitehouse.gov/sites/default/files/microsites/ostp/pcast-health-it-report.pdf>. (accessed 30.04.16).

PMI, 2015. The precision medicine initiative <https://www.whitehouse.gov/precision-medicine> (accessed 30.04.16).

Servant, N., Roméjon, J., Gestraud, P., La Rosa, P., Lucotte, G., Lair, S., et al., 2014. Bioinformatics for precision medicine in oncology: principles and application to the SHIVA clinical trial. Front. Genet. 5, 152.

Shaffer, D.W., Kigin, C.M., Kaput, J.J., Gazelle, G.S., 2002. What is digital medicine? Stud. Health Technol. Inform. 80, 195–204.

Shortliffe, E.H., Blois, M.S., 2014. Biomedical informatics: the science and the pragmatics. In: Shortliffe, E.H., Cimino, J.J. (Eds.), Biomedical Informatics: Computers Applications in Health Care and Biomedicine, fourth ed. Springer-Verlag, New York, NY, p. 15.

Tan, J.J.L., Capozzoli, M., Sato, M., et al., 2014. An integrated lab-on-chip for rapid identification and simultaneous differentiation of tropical pathogens. PloS Negl. Trop. Dis. 8 (7), e3043, <http://dx.doi.org/10.1371/journal.pntd.0003043> Williams, M. (Ed.).

Thuemmler, C., 2015. Health 4.0. Symposium on software defined architectures and networks. <http://thecyberacademy.org/events/sdn/> (accessed 30.04.16).

TMF, 2016. The 40 trends driving the future of medicine. The Medical Futurist <https://gallery.mailchimp.com/5b42ebb547c75ff669a6572d3/files/40_trends_driving_future_of_medicine_Medical_Futurist_Guide.pdf> (accessed 30.04.16).

Topol, E., 2012. Electronic health records and health information technology. The Creative Destruction of Medicine. How Digital Revolution Will Create Better Health Care. Basic Books, New York, NY, p. 145.

U.S. National Institute of Health (NIH), 2010. <https://report.nih.gov/NIHfactsheets/Pdfs/PointofCareDiagnosticTesting(NIBIB).pdf>, <http://www.fda.gov/downloads/MedicalDevices/DeviceRegulationandGuidance/GuidanceDocuments/UCM263366.pdf#page=23> (accessed 30.04.16).

WHO, 2011. mHealth: New horizons for health through mobile technologies: second global survey on eHealth. <http://www.who.int/goe/publications/goe_mhealth_web.pdf> (accessed 30.04.16).

Wicks, P., Chiauzzi, E., 2015. Trust but verify—five approaches to ensure safe medical apps. BMC Med. 13, 205, <http://dx.doi.org/10.1186/s12916-015-0451-z1>.

## CHAPTER 8

# Telemedicine, eHealth and Remote Care Systems

**C.L. Wen**
University of São Paulo, São Paulo, Brazil

## CONTENTS

São Paulo is one of the biggest cities of our Planet and its health system is one of Brazil's best. The quality offered to its citizens probably occupies the middle of the world nations' spectrum. Our Faculty of Medicine's teaching hospital (*Hospital das Clínicas*) is the greatest health center in South America with eight major hospitals that attend about 1.3 million patients per year. This was and is our gazebo to analyze and develop solutions for people's health and disease and we hope that our view of remote care systems may be meaningful for most readers of *Global Health Informatics*.

The concept of remoteness is rather complex. Distance is the aspect that first comes to mind and remoteness always grabs the headlines as, e.g., very recently a patient with supra ventricular tachycardia saved by teledefibrillation up in the Himalayas (Newsletter of the International Society for Telemedicine & eHealth, 2016). However, with the Internet, speed and distances are meaningless and anything unconnected may be considered remote. Indeed, telesolutions in the *Hospital das Clínicas* benefit more patients than in the distant Brazilian rain forests and are also necessary.

Remoteness is not excluded only with web connections. Lack of organization or knowledge are responsible for inaccessibleness in many instances. One of the important goals of Telemedicine in healthcare is multiprofessional integration. In addition the importance of knowledge and awareness cannot be overstressed. Thus the best telehomecare for support and protection of elderly people is useless without competent training to use it.

All these issues must be addressed and the scope of Telemedicine and telehealth are very broad indeed. It is preferable to use both denominations: one to deal with the treatment of patients and the other with health prevention.

Remote care systems is far less a technology problem solving than a useful-ness ensuring.

The Digital Age is a fact. It started on a disputed date of the last century's sec-ond half and caused the fastest and most turbulent transformations since *Homo sapiens* started its journey from the southern tip of Africa. At the Techonomy Conference in Lake Tahoe, California, Eric Schmidt said a few things that are worthwhile to reflect on: "Every two days now we create as much information as we did from the dawn of civilization up until 2003. That's something like five exabytes ($10^{18}$) of data." "The real issue is user-generated content." "I spend most of my time assuming the world is not ready for the technology revolution that will be happening to them soon." (https://www.youtube.com/watch?v=UAcCIsrAq70, 2016).

What is happening may be called an e-revolution and it has transformed almost everything around us, causing a substantial alteration in human behavior. In no time computers migrated from a million desks to a billion pockets and communication is ubiquitous around our Planet. The power of social webs is real and they are determinant in presidential elections and wars. Almost each month cellular phones become more powerful with thou-sands of applications. In this scenario, modern Medicine not only is appall-ingly belated but traditionally very resistant to modifications, in need of urgent creative destruction (Topol, 2013). Here, in Brazil, as almost every-where, the cost of healthcare is unpayable and every year more so. More than 80% of the consultations in the *Hospital das Clínicas* are needless, involving transport, service, and time-consuming costs that could be avoided by Telemedicine programs. Those ailments that really demand treatment more often than not require queueing for months or even more. Informatics and Telemedicine can increase efficiency, practicality, and accessibility to medical services and reduce their costs. Possibly the greatest pressure for changes shall come from the users and not from the personnel accountable for healthcare.

## SOCIAL TRANSFORMATION DUE TO TELETECHNOLOGY

In a great part of modern society the teleconcept is deeply rooted, mainly in urban centers. This is a logical consequence of the diffusion and accessibility to mobile communication accompanied by the increase of computational processing capacity of pocket electronic devices.

The popularization of personal digital assistants started in the 1990s of the last century. The commercial success of its most famous brand, the 3Com's Palm Top, was such that Palm became an independent company from 3Com. The first years of this Century witnessed a great technological upgrade of these computers: the LCD monochromatic screens that only exhibited characters were substituted by color screens with graphic facilities, touch screen interactions, and incorporation of digital cameras ($320 \times 240$ points). Despite the enormous technological improvements, around 2003 the independent devices were condemned to disappear precisely because of their independence. They had to integrate cellular telephony or die. Although the first smartphones appeared in 1994 and evolved as successful models with in-built keyboards like Nokia and Blackberry, as consumers dreams they lasted only a few years. In 2007 Apple launched the first iPhone, the precursor of modern smartphones, as we know them today. This pioneer apparatus integrated all computational qualities of a personal digital assistant with graphic resources, multimedia processing (videos and audios), and an interface for digital cellular phoning with data transmission through 3G. This equipment started a tsunami of behavior changes in communication and data access in the world. Humanity got a connected pocket device with processing capacities.

Although 3G has been available since 2004, Brazil had only 2.5G for cellular phoning until 2007 when iPhone was launched and 3G started to be installed.

Short-distance integration (15–30 m) of different electronic devices by wireless connection became a reality with Bluetooth technology in 2000. A good number of devices adopted it for connection: computers, notebooks, smartphones, tablets, radios, etc. The incorporation of this resource in smartphones and tablets paved the way for their connection with several medical devices.

Everything indicates that with these technological advances added to digital security increase, establishment of social webs, constitution of laws to regulate the digital world, many useful existing services shall go digital offering synchronous or asynchronous connections, where physical distance will not be the crucial factor. Important and rapidly shifting behavioral changes may be foreseen.

Three phases of comparative incorporation speed in society of different technologies may be identified (Anthony, 2012).

1. Insertion phase in the society: when approximately 10% of the population adopts the technology.
2. Maturity phase: 10–40% of the population use the technology.
3. Saturation phase: 40–75% of the population are consumers of the technology.

This tells us that the incorporation speed related to the use of smartphones/tablets technologies is quite fast, with an insertion and maturity phase of less than 5 years. However the saturation perspective is moderated. Thus it seems that the behavioral changes due to the adoption of these gadgets will reach maturity in a short time, certainly until 2020. Examples of behavioral changes due to connectivity in the last decade are Internet banking, webcheck-in, and website income declaration.

In our time important technological leaps are quinquennial and in the coming years they will be even more frequent. Let us remember: iPhone was launched in 2007, WhatsApp in 2009, and iPad in 2010!

## COMPUTER EVOLUTION AND COST DECLINE

It may seem odd, but we live amidst a turmoil of innovations, mostly without perceiving it. The best way to be aware of the transformation immensity is by a short retrospective exercise. Then we may realize that we are living through what was once scientific fiction.

Today we have in our pockets devices that would have been considered supercomputers a few years ago. Let us return to two historical moments of humanity:

1. 1969, Men on the Moon. The computer on board, essential for the mission, was the Apollo Guidance Computer developed by MIT, which used a command interpretation system, called Luminary. Its memory was 64 KB and processing speed, 0.043 MHz. NASA's headquarter used a giant computer, IBM System/360 — 75s Model, for independent calculations and communication between Earth and the Lunar Module. Its cost was US$3.5 million and its size was that of a car. The equipment could realize hundreds of thousand operations per second and its total memory was in the order of megabytes. The programs developed to monitor the environmental data, the spaceship, and the health of the astronauts were the most advanced available.

   2014, Apple's IPhone 6. It had a 64-bit processor A8 ARM, containing approximately 1.6 billion of transistors, operating at 1.4 GHz with a 3.36 billion instructions per second processing power. The clock speed of IPhone 6 is 32,600 times faster than that of the best computer in the Apollo era and can perform 120 million instructions in a bit less than one second (Your smartphone is millions of times more powerful that all of NASA's combined computing in 1969, 2015).

2. February 1997: Human defeated by machine. One of the best known supercomputers is Deep Blue, famous for defeating the chess champion Garry Kasparov. They played 6 games and the champion was defeated in one, drew in two and won three. Deep Blue was considered the 259th most powerful computer in the world with a capacity of 11.38G-FLOPs (Giga floating-points operation per second) and could assess 200 million positions on the chessboard per second.

2014. Samsung Galaxy S5 was equipped with a processor ARM Mali-T628MP6 GPU and had 142G-FLOPs. Today the Tegra K1 mobile processor with GPU 192-core can deliver a peak of 364G-FLOPs (A modern smartphone or a vintage supercomputer: which is more powerful?, 2014).

(Curiosity about FLOP. A simple calculator that is able to do the four basic operations has a processing capacity of 10 FLOPs. In the 1970s the supercomputers reached the capacity of 100 megaFLOPs, equivalent to the processing capacity of a 60 MHz Pentium. Only in the 1980s was one gigaFLOP reached with Seymour's supercomputer Cray-XMP which had the processing capacity of a 350-MHz Pentium II.)

These two striking examples show the unbelievable changes that occur around us without being noticed and raise the question: what can all this processing power hidden in small mobile devices do for Telemedicine, eHealth and for Remote Care Systems efficiency?

## TELEMEDICINE TECHNOLOGY CHEAPENING

Until recently, it was a consensus that Telemedicine was a medical service delivery and its need was determined by distance. The high cost of the equipment and the digital communication between two physical environments for transmissions had to be justified by the results. The investment return was problematic, to say the least. However the present scenerio is completely different and radical changes are expected in the near future. It is probable that the next 5 years will assist the fulfilling of the efficient global web dream. Thus the meaning of Telemedicine must be reconsidered and actualized according to the new sociotechnologic panorama.

The parameters to analyze modern Telemedicine could be the following:

1. The facility and convenience of its service assessing;
2. Uninterrupted availability (24 hours daily and 7 days per week);
3. Service reliability (provided by reference centers and/or with good quality control);
4. Existence of personalized services;
5. Access to multiprofessional treatment;
6. Security;
7. Lower cost;
8. Existence of follow-up resources.

Technological cost factor is changing enormously and very fast. Just consider that about two decades ago the videoconference equipment capable for full duplex transmission at 128 KB, by digital connection (ISDN) with another videoconference device, cost around US$90.000,00. Add to this the price of phone impulses through 64 KB band, charged per minute. Today, in 2016, any US$200 smartphone offers a better transmission through IP connection without additional digital band costs. This means a tumbling in the price by a factor of 450 or more!

The cost change was so enormous that in 5 years videoconferences became incorporated in the daily routine almost everywhere. In this time, some tele-diagnostic and teletherapeutic procedures were standardized and used with increased frequency in high tech countries, such as high-resolution image teleradiology, ultrasound during online videoconferences, tele-ICU (intensive care unit) and teleemergency with remote monitored equipment, telesurgery with holographic visualization and robot arms controlled by a surgeon, among others. With the improvement of digital photography and screen resolution of computers and TV, all services heavily relying on images, dermatology, ophthalmology, pathology, etc., increased their telediagnosis offering.

In 2005 any solution in emergencies that used a color screen palmtop with Bluetooth connection to a computer and a camera glass ($320 \times 240$) cost about US$5000.00. The system was notebook dependent for distance streaming through radiowaves or satellite. At that time, the viability of promising technologies, like WiMax, were discussed. In Brazil, we had only 2.5G band until 2007 when 3D was effectively implanted in the country, and our teleemergency distance by Bluetooth was limited to 30 m. Currently, even the most simple US$200.00 smartphone gets more than 1 megapixel camera and may be connected to a miniaturized filming camera (US$50.00) with much better resolution than those in 2005. Moreover, it covers a much larger area with a much lower transmission cost. In only 10 years, we have a mobile device with resources to access remote data that costs $20\times$ less and is better than its predecessor.

Considering these significant advances, it is necessary to reconsider the concepts and possibilities to provide medical and health services by Telemedicine, eHealth and Remote Care Systems.

## CHANGES IN HEALTH AND RECONSIDERATION OF TELEMEDICINE AND EHEALTH

Change is not a new idea. Famous Greek philosophers like Thales, Heraclitus, and Parmenides tackled the problems of change and permanence. There was no agreement and some philosophical trends favored the ever-changing reality and others the permanent nature of the logical essence. Those who thought that everything seemed to alter and remain at the same time, posed the question: what is the essence that persists during the altera-tions? Twenty-five centuries later, Hegel proposed a synthesis. Consciousness of change can only occur when compared to the permanence. Although the world evolves permanently, the awareness of its evolution cannot occur in real time, but only a posteriori.

This is a fascinating theme. In our days, it is not enough to adapt, we must learn to design the future and prepare for it. In a society in which evolution is accelerated and increasingly more complex, the art of prediction must give way to prospection to meet people's new requirements. It is simpler to pro-duce more than better. Although the quantitative change originates from material solutions, the qualitative one stems from cultural attitudes that result from new ways of interaction, a perspective alteration to things. It is the result of a learning process.

Breaking technology or disruptive innovation is an innovation that helps to create a new network and market value which, eventually, shall disrupt the existing market value network. The term "disruptive" is used in business and technology to describe innovations that improve a product or service in an unexpected way by the market. Although the term disruptive technology is widely used, it may be more convenient to use the term disruptive innova-tion because there are a few technologies that inherently cause major distur-bances; on the contrary, it is the business model allowed by technology that creates a disruptive impact.

The science-fiction series "Star Trek," filmed in the second half of the past millennium, showed an apparatus able to detect every organic problem of a patient: the famous Tricorder (Will The Medical Tricorder From Star Trek Become Real?, 2016). This enormously popular epic story was set in the 23rd and 24th centuries, but it may be said that we, at the beginning of the 21st century, have already a prototype of the Tricorder: the famous smartphone.

Our pocket device has five advantages over the science fiction equipment: it has a built-in communicator that allows communication between two or more people; offers data connectivity to perform collaborative activities and cloud processing; obtains a geolocation system; has great storage capacity and may have a personal assistant for natural voice recognition. Soon we shall have various other refinements, as the appliance will be foldable enough to be wearable, to integrate smart clothes, and be able to enhance resources of biometric identification.

With these changes, the current situation of Telemedicine is at a cross-roads that some economic sectors already faced and overcame a decade ago by means of organized and systematic actions. For its effective consolidation as an efficient solution to increase the spectrum of medical and health services,

Telemedicine and eHealth need to be an extension and expansion of health services, which are provided onsite and moored to hospitals and medical offices.

Unlike the assumption that Telemedicine could compete with traditional medical activities, make them more inhuman or cause a devaluation of professional services, Telemedicine is a proposal for innovation that can improve the systematic of implementation and provision of services. It can increase the efficiency, may expand users' coverage period, and is able to create new services that can improve the integrated chain of processes that by conventional methods would be impractical, as, e.g., the incorporation of customizable services to fit the profile of each individual.

Telemedicine is a disruptive innovation based on interactive technologies in restructuring health processes, in monitoring the speed of the changes of an increasingly globalized world, with ever greater and deeper knowledge.

The advances of electronic component miniaturization creates and disseminates smartphones with ever-greater processing capabilities to increasing number of people. This fact indicates these devices as the best choice for social assistance tools. If associated with special medical devices, they can give support to clinical propaedeutic and diagnosis.

What was considered until 15 years ago as procedures or specialized exams for diagnosis support, probably until 2021 will be considered as propaedeutic standard and mandatory resources. This disruption in the traditional practice of medical services will bring other major changes, as, e.g., the possibility to expand sites designated for disease and/or healthcare, hospitals and medical offices, extending them to homes and communities. With the cheapening and increasing of the access quality to improved connectivity, we will have the possibility to offer health services beyond diagnosis and treatment of diseases: health monitoring services and increased resources to achieve early diagnosis of diseases.

Probably many services will no longer be dependent on humans but on automatic electronic personal assistants, with natural language recognition capability, identification of the user, storage of its problems and utilization of his profile to do any research or/and needed processing in a health-cloud.

## MODERN TELEMEDICINE ACTIVITIES

Although Telemedicine is still associated with the concept of social assistance service provisions, it is time to understand it as the integration of a number of processes with the use of teletechnologies. It participates in distributed processing and is essential for structuring integrated services with a focus on

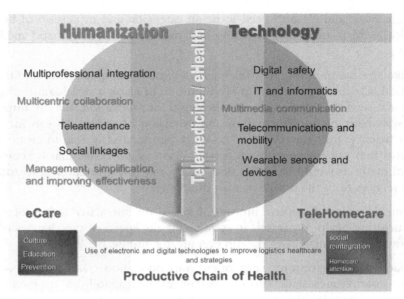

**FIGURE 8.1**

Illustration of the productive chain of health.

health management, diagnoses and treatment streamlining and increased access to recovery and social reintegration services. As well as Medicine, Telemedicine and eHealth operate in three segments: (1) teleassistance and epidemiological telesurveillance; (2) interactive teleeducation, and (3) multicentric research network (telescience).

In the Integrated Health Chain context, it is possible to group Telemedicine into three tracks: (1) eCare: focused on education, promoting life quality and helping to diagnose early morbid processes; (2) teleassistance: focused on optimization of disease treatment logistics; (3) telehomecare (Fig. 8.1).

eCare is an area that will likely have significant growth in the near future, because of the importance to recognize diseases in their early stages for obvious benefit of the patient and also the economy. For that purpose, the field is enormous and widespread technologies in the population may have an important role in the hands of professionals. In eCare, we can work on three axes:

1. Knowledge increase about health in the schooling system and promotion of healthy habits in the early formative stages of the future citizens. This period is most appropriate to develop a whole series of practical learning with behavioral change. It is possible to correlate priority health issues to the curricular grids and develop activities

unraveling the mysteries of the human body and its health maintenance. In Brazil we elaborated the Young Doctor Project (YDP) that promotes health in schools through a creative learning, bringing together undergraduates from the University and high school students. We will cover this topic in more detail later in this chapter.

2. Interactive guidance on health platforms. These are interactive environments with specially produced videos (and may have 3D computer graphics of the human body), educational games, situation simulators, group dynamics, video-meetings, and use of virtual reality resources for additional explanations to patients and their families, with a focus on understanding and strengthening healthy habits. On this platform, health communication and interactivity are products especially designed to facilitate the understanding of the various specific issues that may also increase the patient's commitment to the prescribed treatment. Some surveys show that in Brazil, 50% of therapeutic failures are due to misunderstanding or abandonment of treatments by patients. Thus this work can enhance treatments through standardized guidelines delivery. An example of the efficient use of this resource is the standardized and objective guidance designed to increase the adherence of patients to treatment of systemic arterial hypertension in the Heart Institute's ambulatory of the *Hospital das Clínicas*.

3. The importance of diagnosis in the early phase of the diseases cannot be overemphasized. To achieve success, the actions need to be done consistently, systematically, in large regions and close to the population. This means most of the time outside of the hospital environment and generally far away from services with infrastructure for the diagnosis. It is necessary to provide training courses addressed to specific professional groups, empowering them to care about diseases, health promotion guidelines, and use of low-cost devices for data submission of people with suspected diseases to authorized services with digital screening for diagnosis. Example: teaching photograph skills to a beauty salon hairdresser to document skin lesions by means of a smartphone could be a way for early recognition of skin cancer; to pedicures for the identification of diabetic foot lesions or foot plant melanoma; tattoo artists or masseurs could identify skin lesions in other parts of the body (Chao et al., 2003a; Oliveira et al., 2002). Already with the current resources of Internet video transmission, it is possible to transform beauty salons into health broadcast centers (Health's NetFlix). Our experience involves training of professionals about healthy living with the Sun, suspicion of early lesions of skin cancer, prevention of sexually transmitted diseases, and sanitation of beauty salon instruments (Vieira Júnior and Chao, 2015).

## REMOTE CARE SYSTEM

To define Remote Care Systems is not an easy task. Probably the most associated idea with this term are physical devices appropriated to capture and transmit patient data. This perception is poor because modern smartphones are able to exchange data for cloud processing and some of its accessories turn them into high-tech Remote Care System devices.

Remote Care Systems may be organized into five groups:

1. Health monitoring applications (software) for smartphones and tablet, which allows sending individual data for monitoring health situation of an individual (Are Digital Health apps future drugstore products?, 2013). The popularity and the increasing possibility to incorporate resources of personal assistants (software) with artificial intelligence in smartphones integrated with GPS and collaborative applications to access services and/or instructions based on videos will be responsible for new forms of interaction with patients and between groups of people. These features will be important allies to promote habit changes, increased adherence to treatments, continued explanations of medical guidelines and/or monitoring the correct use of medicines, and, eventually, automated dispatch of drugs for chronic use, mainly when there are apps developed for specific situations of chronic diseases or special health conditions, like diabetes, hypertension, obesity, and others.

2. Mixed resources of electronic apparatus and apps: portable equipment able to perform various examinations or capture biological signals of patients. For example, tele-ECG, portable ultrasound, digital oximeter, dermatoscope, among others.

3. Cloud productivity software and online communication apparatus that enable interaction between people by means of smartphones and computers. In this category there are several well-known software types, which although used for other purposes, with data security and confidentiality assurance, to some extent could be used for Telemedicine assistance purposes. Among the tools, we could quote WhatsApp (online communication with organization resource for groups and video calls), Facebook (social network with resources to organize closed online communications, data sharing, video calls, virtual reality, image recognizer, and 360 degree viewer), Google and Microsoft productivity tools (Office 365).

4. Monitoring systems using accessories for general use as a complement for standard smartphone resources, such as smart watches, holographic cameras, 360 degree cameras, adapters for virtual and immersive reality, intelligent bracelets, etc.

5. Games using virtual reality and augmented reality can be employed for monitoring various situations of patients, e.g., in the case of people with motor and cognitive independence limitations. These tools also allow distance training and open up the possibility of prescribing activities and monitoring exercise quality performed at home.

## DEVICES TO SUPPORT TELEASSISTANCE

Smartphone apps for medical purposes are multiplying quickly; new portable electronic devices to capture biological signals are constantly released and their power increases. Soon, many services that are still considered specialized activity or propaedeutic procedure will be mandatory and part of routine exams. Examples of medical equipment are as follows:

1. Portable ultrasound that communicates with tablet or smartphone. It is estimated that these devices, simple and light, have image quality equal to or even 10% higher when compared to the ultrasound apparatus used in the second half of the 1990s by Japanese physicians to stage liver cancers. The models with concave and linear transducers, besides performing organ image scans, may be used to guide other medical procedures, such as needle biopsies. The current commercial models have 64 piezoelectric crystals for quality imaging that allow examining organ structures, helping propaedeutic techniques, like abdominal palpation, cardiac, gestational and joint assessments, and also serve for guidance on many procedures. This is an important advance for bedside Medicine in distant or isolated regions (Ferreira et al., 2015; Helland et al., 2016). Of course, the possibility to record still images or video sequences enables us to obtain expert opinions from any distance. Thus even though these apparatus do not have comparable quality to the most modern ultrasound equipment, they offer ease portability, widespread use, and serve to perform initial specialized assessments almost everywhere (Arbeille et al., 2016; Adhikari et al., 2014; Marshburn et al., 2014).

2. ECG equipment for smartphones with individual electrodes, integrated smartwatches or smart clothes or coupled with cell phone covers. These devices allow ECG registry, storing and shipment to cloud to request specialized reports. The initial models of the tele-ECG apparatus used modems for data transmission by telephone line or Internet transmission when coupled with microcomputers. The second phase showed independent equipment that sent data directly by a mobile device and now we have the smartphones for ECG recording and transmission. Probably, very soon the registers may be

sent to a Health Cloud with a database and system for checking the ECGs automatically, as a preassessment.

3. Appliances and adapters for dermatoscopy and external ocular photography using a smartphone to register dermatological lesions and/or ophthalmologic alterations. Integrated with pattern recognition software, automatic screening can be obtained. The availability of apps for smartphone photography and image recognition is a reality. Although digital photographs and dermatoscopy images may be identified, it is very important to add correct clinical data to secure diagnosis accuracy. The use of adapters for pictures can be a cheap way to follow up treatments of chronic wounds, systemic diseases, differential diagnosis of red eye, etc. (Chao et al., 2003b; Trindade et al., 2008; Soirefmann et al., 2013; Taleb et al., 2005).

4. Ophthalmoscopes coupled with smartphone. These devices allow background examination of the eye with video recording and/or still images for assistance purposes by the physician who is attending the patient or sending them for specialized consultation or monitoring of some systemic diseases through fundus examination. In addition to commercial devices, there are also adaptations using 3D printers for low-cost funduscopic exam (Sreelatha and Ramesh, 2016; Ye et al., 2014).

5. Otoscope and adapters for rigid or flexible endoscope to ear and throat exams, recordings, and second opinion consultations.

6. Portable optical microscopes for smartphones, with magnification capacity between 60 and 200. These low-cost equipments allow various exams and are especially useful in detection of fungal hyphae in scraped lesions, parasites in feces, and other situations that do not require special coloring (Switz et al., 2014; Breslauer et al., 2009).

7. Colposcope for smartphones. Accessories for uterine cervix exam with still image capture or video recording. Very useful tools for prevention of cervical cancer campaigns (Spitzer, 2015; Catarino et al., 2015).

8. Digital stethoscope. They may be coupled with smartphones for recording auscultation. There are also commercially available small cheap devices that can perform functions similar to those of a stethoscope. They are being used by the population for purposes of self-care, e.g., during physical exercise to monitor heartbeat and pregnancy to exam the fetal heart condition (Lakhe et al., 2016; Kevat et al., 2015).

9. Digital oximeter for smartphone.

10. Electroencephalogram smartphone-based device for registration of brainwave patterns in graphic form.

11. Doppler Fetal Monitor for fetal auscultation by smartphone.

12. Spirometer based on smartphones.
13. Glucometer based on smartphones.
14. Biochips to carry out 20–100 laboratory tests with a drop of blood and presenting the results in a smartphone (Reddy et al., 2016; Huang et al., 2016).
15. Other equipment and apps are created almost each month, perhaps better to say each week, enriching and improving remote care in the world.

This range of devices indicates that soon there will be a profound change in the form of patient assistance and home care in the broad sense will be more and more a reality in most parts of our Planet.

In addition to assistive devices to diagnosis, advances that occurred in cloud processing and the emergence of the so-called digital safe-rooms will lead to feasible interconsultation services and specialized second opinions with digital data record, in an agile and safe manner. Online meeting rooms with a large amount of participants, use of smartphone-based virtual reality, social medias, cloud-based productivity tools, among others, will substantially change the interaction forms, clinical meetings, and collaborative work.

## ATTENTION TO RESIDENCE AND TELEHOMECARE

The world population is aging and more and more people will need medical assistance. The bill will not reduce if Medicine continues to be practiced only in doctors' offices and hospitals. However the scenario may change according to the analysis of a group of specialists of Ernst & Young International Consultancy, which subsidizes companies interested in the market trends of Medicine. According to their research, an important part of medical care in the future will be made in the so-called "third place" (the other two are the hospital and the doctor's Office).

This "third place" could be, e.g., the home of patients with chronic diseases such as diabetes, obesity, or respiratory ailments, or any other place where they are. The main innovations in healthcare and eHealth will come from technologies that facilitate remote assistance, such as apps for tablets and mobile phones to remember the time of taking drugs, for example. It is estimated that 75% of healthcare costs come from chronic diseases and the number is likely to increase. These patients do not need to be in the hospital but require medical constant follow-up, which may be offered in new ways. This will cause a behavior change of doctors and patients (The future of health insurance, 2016; Changing behaviors represents the single biggest opportunity to improve health outcomes, 2012).

The rise in life expectancy and average age in the world, with increasing population of elderly persons; habit changes leading to overweight/obesity, chronic diseases and degenerative metabolic diseases, physical disability and neoplastic diseases; the risk of pandemics by increased globalization, travels, and displacement, among other features, are modifying disease profiles. They also show that in addition to the efforts to provide treatment of diseases, it is necessary to incorporate new strategies and services for disease prevention and health promotion. The extent of possible health actions indicates the need to create new strategies with multiprofessional and integrated approaches, to organize lines of health attention to solve problems and actions, in order to minimize the resurgence and avoid exacerbation of diseases, especially chronic ailments.

In various parts of the world there are Health Policy Organizations centered on factors that affect the health of the population and diseases that are of major occurrence, studying possibilities of local resolution through less specialized infrastructure. Their concern is with primary care proposing to avoid worsening of the conditions, unnecessary displacements, and the emergence of a chain of consequences with negative impact throughout the health system. Although individual approaches to diseases in primary healthcare are less complicated, as a whole, a community approach is quite complex due to logistic organization, distribution of health resources, and maintenance of professional actions in an integrated manner.

The systems for synchronous home teleassistance will become popular and more reliable with the use of dedicated high-quality cameras together with 180-degree or 360-degree viewing cameras for environmental situation evaluation. These resources are very useful for management of elderly people and those with physical disabilities (Huang and Chan, 2014; Zhang et al., 2013).

## COMMUNICATION IN HEALTH—INTERACTIVE AND OBJECTIVE

Efficient communication and recreation is one of the powerful tools for health promotion and increase of patients' adherence to treatment. In this context, it is important that communications should be produced in a professional manner taking into account the sociocultural archetypes of the target audience.

The advance of graphic capabilities accelerated enormously in recent years with the popularity of 3D computer graphics, high-quality digital objects in computer systems, virtual reality, and interactive games.

The Virtual Human Project started in the Discipline of Telemedicine of the Department of Pathology of USP's Faculty of Medicine (DTM-FMUSP)

**FIGURE 8.2**

Images from the Virtual Human project.

in 2003. It involves the use of 3D graphic communication associated with dynamic resources to produce thematic videos (Dynamic and Directed Communication). One of its features is the ability to concentrate large amounts of scientific information (between 30 and 100 pages of scientific descriptive texts) in a video sequence of 1−3 minutes. It is also able to convey thematic information in an organized and fluid manner according to a predetermined pedagogical script. The sequences of Virtual Human (VH) may be called Educational Learning Objects and can reduce up to 70% the Professor's time during the teaching-learning process in Anatomy, Biomechanics, Biomolecular Medicine, Physiology, Physiopathology, and medical skills/procedures. The great timesaving gives more time to promote interactions, develop observation skills, critical analysis, reasoning, and to promote practical contextualization of the subjects under discussion.

The productions of the VH Project result from the integrated work of three professional groups: teachers and researchers in interactive education of telemedicine, 3D computer graphics designers and specialized professionals in the matters under construction. Together, they build structures through the modeling of a mesh of 3D computer graphics (Chao, 2016) (Fig. 8.2).

Being entirely in digital format, the VH collection is available over the Internet and can be distributed to anywhere in the country or the world. With minor adjustments, every structure modeled in 3D may be used for the production of physical structures by means of 3D printers; be transformed into digital objects to be installed in smartphones and tablets with interaction resources; or be used to produce applications in virtual reality or augmented reality.

The contents of the VH, considering their condition of educational learning objects, follow a standardized systematics, which aims to guarantee production quality. Depending on the complexity, a thematic sequence can take from 2−8 months to be fully produced, depending on the level of detail of the anatomical structures, complexity of physiological animations or actions, and the need for supporting materials (scientific literature, videos and images, models, slides, etc.).

Over the 13 years of existence, the VH Project produced:

1. 500 thematic sequences for the areas of Medicine, Nursing, Phonoaudiology, and Odontology;
2. 300 digital structures by 3D printer;
3. 10 interactive digital objects for use in smartphones and tablets;
4. Assistance structures, like skull pieces, articulations, and others, produced by 3D reconstruction from images of computed tomography (CT) and magnetic nuclear resonance (MNR).

Some details of these productions:

1. Thematic sequences of VH: These videos, with length between 1 and 3 minutes, facilitate the exposure and help to learn the most relevant points of complex subjects, quickly and easily, integrating anatomy, physiology, physiopathology, histology, and, if necessary, biomolecular mechanisms (Camargo et al., 2011).
2. Digital structures by 3D printer: Digital files generated from the models of VH for physical structures, with realistic detailing. These structures, in addition to serving as a first step toward studying Anatomy (descriptive, topographic, and spatial), present as a highlight the possibility of learning correlations between Anatomy and methods of diagnostic imaging (X-ray, ultrasound, CT, MNR etc.), as well as facilitating understanding correlations between Anatomy and Clinical Propaedeutic and helping to develop investigative reasoning (Fig. 8.3).

   The cheapening of 3D printing associated with the improvement of their quality opens the possibility of high-quality educational product distribution on the Internet. The fact that the structures in question are produced from materials such as plastic, plaster, or resin permits their use outside the anatomy laboratories, anywhere, in clinics, hospitals, libraries, or even at home.

   The digital files for the production of such structures can be sent over the Internet, enabling the production of physical models in any

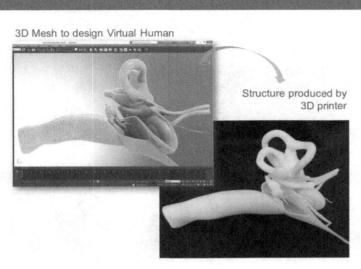

**FIGURE 8.3**

An example showing a 3D structure produced from a computer graphic model.

institution with 3D printers and their customization according to specific didactic needs.

3. Interactive digital objects: Equivalent to files used for production by 3D printer. However, in this case, the available structures may be employed as applications installed on computers and mobile devices, with resources for addition and subtraction of structures, application of transparencies, expansion and, sometimes, with dynamic features. These objects, when associated with VH animation sequences and structures produced by 3D printer, may form a chain that facilitates learning by joining interactivity with dynamics and management of physical structures.

4. Assistance structures: In addition to educational productions, the team of designers of VH is able to produce pathological organs from images obtained from virtual autopsy in the FMUSP by a 7-T MNR device. They also created tailor-made prostheses and anatomical structures of CT-scanned patients for study and planning of complex surgical interventions (Fig. 8.4).

The modernization and creation of health spaces; the use of virtual reality; the incorporation of interactive features that enable the visualization of Anatomy and Physiology; and the use of 3D printers, as instruments for production of physical objects, certainly facilitate understanding and stimulate learning.

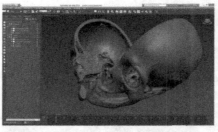

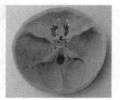

Radiology – CT - Dicom images

Structures re-created from Dicom images for surgery planning

**FIGURE 8.4**
Structures re-created from Dicom images for surgery planning.

## EDUCATION, HEALTH PROMOTION, AND SCREENING

Early learning about health and disease prevention is one of the best ways to promote habit changes in a society and educate citizens who have knowledge of their body and care for it.

Education is a complex process and the use of computational resources, interactive technologies, and mobile communication undoubtedly help learning. More and more computers participate in our daily life and so it is natural that education uses the advantages of this technology and applies its potential as an important tool in the learning process.

In developed countries, e.g., the museums of Science are important alternative learning tools. A comprehensive program disclosing various fields of culture and science gives the student a broader vision of the world in which he lives. As Falk and Dierking state, the overwhelming majority of learning outside of school is by choice driven by the apprentice's needs and interests (Dierking and Falk, 2005). This is as important as the conventional schooling, with the advantage of interacting between individuals and enhancing socialization.

The Young Doctor Program (YDP) devised by the DTM-FMUSP is an uninterrupted activity since 2007 and is successful. Contrary to what the name may suggest, the YDP does not mean a young physician but a young person with a good knowledge in some specific subject able to help a community. "Doctor" in Brazil has a broader meaning than in some other nations. It is a multidisciplinary activity, which uses Telemedicine capabilities, interactive education and VH, with the purpose of encouraging high school students, professionals from Basic Health Units, and undergraduates from the universities to perform cooperative work through sustained action with the objective of promoting health and better quality of life of communities in need.

YDP is an opportunity to exercise citizenship, university extension activity, and scientific initiation, with practical application of knowledge obtained in classrooms under the supervision of teachers. It tries to encourage the students to understand the characteristics of basic attention to health and the needs of underprivileged communities. The interaction of students from all kinds of professions helps to promote the global health of selected communities. For high school students the YDP represents a chance for digital inclusion and learning about health, through the Extension Courses offered by the university. It also enables the development of a social role in their own community, integration with the Basic Health Units of their region and acquaintance with the city's health infrastructure.

The YDP is an action of the university to the society, which establishes a process of social commitment in various regions and professional segments. The themes are selected and implemented according to the needs and local priorities of the communities, creating a bond of responsibility and motivation.

The involvement of student strength is important. It accelerates their professional qualifications, and their participation in social well-being promotion is a welcome action by the Government because it enhances the formation of the future citizen. The youngsters have the opportunity to be acquainted with different Brazilian realities and do some work to promote improvements in the quality of life of the visited populations, under supervision of qualified teachers.

The idea starts from the supposition that scientific knowledge related to communication techniques and organization of practical activities can promote learning of health knowledge through contextualization. The use of interactive Internet technologies can be an efficient means to spread social education in health, more broadly, with the use of multiple media to engage the target audience, establishing a model of nonformal education suited to the modern world. Knowledge transmission occurs through a fluid communication, with structured scripts in an accessible language to facilitate information understanding and learning construction by association of cognitive information (theoretical) with static and/or dynamic images (VH Project), group interaction (Team Based Learning), development of thematic projects (Project Based Learning) and by carrying out practical activities.

Scientific knowledge in the area of health is still little understood by a large part of the population. Evidence of this, e.g., is the high rate of violence caused by the consumption of substances harmful to health, such as alcohol (Pereira et al., 2015), drugs, and cigarettes. Awareness of what needs to be done to keep the body healthy and have a good social conviviality is one of the alternatives sought to decrease rates of sickness and death caused by these substances. The use of multiple ways to engage the target audience is a proposal of a differentiated model of nonformal education: the use of increasingly interactive technologies to inform.

One of the important purposes of this project is the dissemination of educational material exploring the potential of students to learn to learn, to develop critical opinion, to research and explore its multiple functionalities. It is also important to take new technologies for training young people and use them as allies in the teaching/learning process.

To allow students to manipulate learning objects (VH), formation of virtual communities to discuss issues under the guidance of a tutor and organization of activities in their communities are important steps in this project. They increase the participants' interaction with the subjects under discussion, the degree of motivation and will offer them the experience of the proposals by means of resources which, in turn, may promote learning that will make "a relatively permanent change in behavior that results from experience" (Antunes, 2002).

The availability of good instructional materials and guidance to foster curiosity are represented by 3D videos of computer graphics (units of knowledge), fiction videos (dramaturgy), scripts, posters, and anatomical structures produced by the 3D printer. They address issues on drugs, alcohol, smoking, "know your body," sexuality, contraception, prevention of sexually transmitted diseases, hand washing, infections by H1N1, Zika, Chikungunya, Dengue, and others.

The set of videos of VH (knowledge units) and fiction approximates the public and the health issues through a simple and accessible language. In addition, it awakens the public interest and curiosity to browse and to get more information about what was seen.

Thus this action aims to deploy and show that the structuring of Health Cultural Spaces for teaching can be a propagator of culture and life quality, while the individual experiences reality and is motivated to seek knowledge and to share it with others. The role of communication in this context is to structure the way to disseminate information and build cultural values capable to perpetuate over time.

Using the current technology, it is possible to develop a model of the relationship between students and health professionals with the population, empowering a dynamics in which the focus is the promotion of health education with encouragement of life quality through a more contextualized and humanized approach. With this, the work may be catenated (Productive Chain of Health) to generate positive cycles in health promotion and well-being of a community in addition to reducing health costs and develop awareness in schools to introduce life quality promotion and educational strategies to reduce disease risks.

Interactive learning environments approach the public with issues related to health, through a simple and accessible language to facilitate understanding of these subjects, motivate, and promote a greater spread of useful knowledge to the communities. Hopefully, it also raises the interest and the curiosity of young people so that they seek or obtain more information about what was shown to them.

To promote sustainability in the implementation of actions, a strong incentive and involvement of the Municipal Education and Health departments is mandatory.

## VIRTUAL REALITY PERSPECTIVES

The popularity of virtual reality will bring a new social situation. With the technological sophistications and increased data transmission capabilities, the digital reality approaches reality and it is quite problematic to separate both. On the other hand, this may be advantageous, since several simulations of skills and activities that can promote rehabilitation (Kavamoto et al., 2005) become possible. However, it also may generate negative situations, where effort will be needed to avoid the mixing of the two realities; not to mention a multitude of aspects related to ethics and responsibilities.

The immersive reality using smartphone is becoming too common and it integrates virtual reality, holographic vision, and spatial control with movements. This resource is enabling increasingly realistic navigations and interactions. Its use, in addition to entertainment and educational purposes, has potential advantages in rehabilitation, equilibrium training and motor coordination, as well as mental health, e.g., desensitization of panic reactions. When coupled to video equipment with a joystick using accelerometers or motion recognition, the system of virtual games can be applied to training and performance monitoring of motion recovery.

## 3D PRINTERS AND THEIR INSERTION IN THE EHEALTH WORLD

Although it has been a developing technology since the 1980s, it is only in this decade that 3D printing technology has become popular, and by all indications will have a transforming role in society. These equipments opened the possibility for physical production of custom-designed structures in the digital world. The most popular models of 3D printer, which can be bought in shops and supermarkets, use basically plastic filaments (PLA and ABS) and are able to produce structures with sufficient quality for educational purposes.

There are several applications of 3D printers; in the area of health what is best known today is the production of tailor-made prostheses (Park et al., 2016), production of anatomical structures for educational purposes and/or complex surgeries (Zeng et al., 2016; Muelleman et al., 2016) and production of guides for minimally invasive surgeries. In the near future, "4D printers" will occupy the limelight performing bioprintings (Zhang et al., 2016) primarily intended for reparative surgeries. At this time, 3D printers are a kind of factory of smartphone adapters that can be materialized at distances. This may be advantageous to distant locations. For example, the user can download an adapter to make his mobile device or tablet appropriated for teletransmissions or to perform some procedure more accurately (3D printable microscope for home or lab, 2016; UC Berkeley showcases CellScope with 3D printed iPhone stand, 2012). This concept enables various resources. Some odd applications are in progress with 3D printers, like production of food and personalized drugs (Yang et al., 2015; Norman et al., 2016; Jonathan and Karim, 2016).

The main characteristic of 3D printers is that its coordinating information can be transmitted through a digital network. This opens up the possibility of production, management, and large-scale distribution.

## PERSPECTIVES—NEXT STAGES—GRAPHENE AND BIOCHIPS

In these next 5 years, we shall have important advances as a result of new releases of electronics based on Graphene technology. It has the potential to popularize malleable and folding equipment, new technologies for batteries that allow 2000 recharges (practically 20–30 years), the expansion of 3D printers for different applications, the advancement of electronic gadgets that are becoming unbreakable and water resistant, and the expansion of LiFi (Light Fidelity)—Internet connectivity by light and advent of the 5G band.

The popularization of Graphene technology opens the possibility of wearable technology systems. In addition, the clothes may be connected and have flexible devices and sensors in contact with the body, transmitting parameters and various data to help health management.

All these changes that will occur in society as a whole indicate that the medical and health areas demand urgent modernization. They badly need new educational methods, organization of technological actualization programs for professionals already active on the market to create specialized groups in Telemedicine and eHealth innovation, organize new methods of hybrid propaedeutic, and elaborate new aspects of ethics, responsibility, and digital security in health.

## FUTURE OF TELEMEDICINE

In summary, it can be said that Telemedicine and eHealth will be endless. The tendency is a consolidation within the social processes of humanity. Some experts believe that soon there will not be any sense to call a specific area Telemedicine, because the technological resources will be completely incorporated into the day-to-day activities of the various medical specialties. Nevertheless, one can also envisage that progress and changes are going to be ever more accelerated with the emergence of new technologies, and we shall need dedicated professionals to evaluate and to integrate them with the praxis of Medicine.

Managing sustainability, promoting renewal in the processes, following the changes of the regulatory aspects of the digital world, specifying its limitations, as well as research and encouragement of the teaching of new propaedeutic methods are the premises of Telemedicine and eHealth professionals.

## Acknowledgment

Our special thanks to György Miklós Böhm, Emeritus Professor of the FMUSP, for his dedication to discussing each of the topics covered in this chapter and help to review and translate the text.

## References

3D printable microscope for home or lab. Retrieved July 20, 2016 from <http://www.instructables.com/id/3D-printable-microscope-for-home-or-lab/>.

Adhikari, S., Blaivas, M., Lyon, M., Shiver, S., 2014. Transfer of real-time ultrasound video of FAST examinations from a simulated disaster scene via a mobile phone. Prehosp. Disaster Med. 29 (3), 290–293, <http://dx.doi.org/10.1017/S1049023X14000375>.

Anthony, S., May 9, 2012. Smartphones set to become the fastest spreading technology in human history, May 9, 2012. Retrieved July 20, 216 from <http://www.extremetech.com/computing/129058-smartphones-set-to-become-the-fastest-spreading-technology-in-human-history>.

Antunes, C., 2002. Novas maneiras de ensinar, novas formas de aprender. Artmed, Porto Alegre.

Arbeille, P., Zuj, K., Saccomandi, A., Ruiz, J., Andre, E., de la Porte, C., et al., 2016. Teleoperated echograph and probe transducer for remote ultrasound investigation on isolated patients (study of 100 cases). Telemed. J. E Health 22 (7), 599–607, <http://dx.doi.org/10.1089/tmj.2015.0186>.

Are Digital Health apps future drugstore products?, April 17, 2013. Retrieved July 20, 2016 from <http://www.digitalforhealth.com/blogs/category/iphone>.

Breslauer, D.N., Maamari, R.N., Switz, N.A., Lam, W.A., Fletcher, D.A., 2009. Mobile phone based clinical microscopy for global health applications. PLoS One 4, e6320.

Camargo, L.B., Aldrigui, J.M., Imparato, J.C.P., Mendes, F.M., Chao, L.W., Bönecker, M., et al., 2011. E-learning used in a training course on atraumatic restorative treatment (ART) for Brazilian dentists. J. Dent. Educ. 75 (10), 1396−1401.

Catarino, R., Vassilakos, P., Scaringella, S., Undurraga-Malinverno, M., Meyer-Hamme, U., Ricard-Gauthier, D., et al., 2015. Smartphone use for cervical cancer screening in low-resource countries: a pilot study conducted in Madagascar. PLoS One 10 (7), e0134309, <http://dx.doi.org/10.1371/journal.pone.0134309> eCollection 2015.

Changing behaviors represents the single biggest opportunity to improve health outcomes, 2012. Retrieved July 20, 2016 from <http://www.ey.com/GL/en/Industries/Life-Sciences/Progressions-2012---Health-care-everywhere---Overview>.

Chao, L.W., 2016. Virtual human 3D: the integration of computer graphics, 3D printing and virtual reality for anatomy, physiology and pathophysiology learning. Rev. Grad. 1 (1), 7−15, ISSN 2525-376X. Retrieved from <http://www.revistas.usp.br/gradmais/issue/view/8826/>, <http://dx.doi.org/10.11606/issn.2525-376X.v1i1p7-15>.

Chao, L.W., Enokihara, M.Y., Silveira, P.S.P., Gomes, S.R., Böhm, G.M., 2003a. Telemedicine model for training non-medical persons in the early recognition of melanoma. J. Telemed. Telecare 9 (Suppl. 1), 4−7, <http://dx.doi.org/10.1258/1357633303322196141>.

Chao, L.W., Cestari, T.F., Bakos, L., Oliveira, M.R., Miot, H.A., Böhm, G.M., 2003b. Evaluation of an Internet-based tedermatology system. J. Telemed. Telecare 9 (S1), 9−12, <http://dx.doi.org/10.1258/1357633303322196169>.

Dierking, L.D., Falk, J.H., 2005. Using the contextual model of learning to understand visitor learning from a science center exhibition. Sci. Educ. 89 (5), Version of Record online.

Ferreira, A.C., O'Mahony, E., Oliani, A.H., Araujo Júnior, E., da Silva Costa, F., 2015. Teleultrasound: historical perspective and clinical application. Int. J. Telemed. Appl. 306259, <http://dx.doi.org/10.1155/2015/306259>.

The future of health insurance. Retrieved July 20, 2016 from <http://www.ey.com/GL/en/Industries/Financial-Services/Insurance/ey-health-insurer-of-the-future>.

Helland, G., Gaspari, R., Licciardo, S., Sanseverino, A., Torres, U., Emhoff, T., et al., 2016. Comparison of four views to single view ultrasound protocols to identify clinically significant pneumothorax. Acad. Emerg. Med. 23, 1170−1175, <http://dx.doi.org/10.1111/acem.13054>.

Huang, C.N., Chan, C.T., 2014. A ZigBee-based location-aware fall detection system for improving elderly telecare. Int. J. Environ. Res. Public Health 11 (4), 4233−4248, <http://dx.doi.org/10.3390/ijerph110404233>.

Huang, N.L., Ye, L., Schneider, M.E., Du, Y.X., Xu, Y.H., Fan, L.B., et al., 2016. Development of a novel protein biochip enabling validation of immunological assays and detection of serum IgG and IgM antibodies against *Treponema pallidum* pathogens in the patients with syphilis. Biosens. Bioelectron 75, 465−471, <http://dx.doi.org/10.1016/j.bios.2015.08.036> Epub September 5, 2015.

Jonathan, G., Karim, A., 2016. 3D printing in pharmaceutics: a new tool for designing customized drug delivery systems. Int. J. Pharm. 499 (1−2), 376−394, <http://dx.doi.org/10.1016/j.ijpharm.2015.12.071> Epub January 3, 2016. Review.

Kavamoto, C.A., Chao, L.W., Battistella, L.R., Böhm, G.M., 2005. A Brazilian model of distance education in physical medicine and rehabilitation based on videoconferencing and internet learning. J. Telemed. Telecare 11 (S1), 80–82, <http://dx.doi.org/10.1258/1357633054461949>.

Kevat, A.C., Dawson, J., Davis, P.G., Kamlin, C.O., 2015. Evaluation of a digital stethoscope and smart device technology for assessment of heart rate in the newborn infant. Arch. Dis. Child Fetal Neonatal Ed. 100 (6), F562–F563, <http://dx.doi.org/10.1136/archdischild-2015-308639> Epub July 10, 2015. No abstract available.

Lakhe, A., Sodhi, I., Warrier, J., Sinha, V., 2016. Development of digital stethoscope for telemedicine. J. Med. Eng. Technol. 40 (1), 20–24, <http://dx.doi.org/10.3109/03091902.2015.1116633> Epub January 5, 2016.

Marshburn, T.H., Hadfield, C.A., Sargsyan, A.E., Garcia, K., Ebert, D., Dulchavsky, S.A., 2014. New heights in ultrasound: first report of spinal ultrasound from the international space station. J. Emerg. Med. 46 (1), 61–70, <http://dx.doi.org/10.1016/j.jemermed.2013.08.001>.

A modern smartphone or a vintage supercomputer: which is more powerful?, June 14, 2014. Retrieved July 20, 2016 from <http://www.phonearena.com/news/A-modern-smartphone-or-a-vintage-supercomputer-which-is-more-powerful_id57149>.

Muelleman, T.J., Peterson, J., Chowdhury, N.I., Gorup, J., Camarata, P., Lin, J., 2016. Individualized surgical approach planning for petroclival tumors using a 3D printer. J. Neurol. Surg. B Skull Base 77 (3), 243–248, <http://dx.doi.org/10.1055/s-0035-1566253> Epub November 3, 2015.

Newsletter of the International Society for Telemedicine & eHealth, quarterly publication, July 2016.

Norman, J., Madurawe, R.D., Moore, C.M., Khan, M.A., Khairuzzaman, A., 2016. A new chapter in pharmaceutical manufacturing: 3D-printed drug products. Adv Drug Deliv Rev. pii: S0169-409X(16)30077-1. <http://dx.doi.org/10.1016/j.addr.2016.03.001>. [Epub ahead of print] Review.

Oliveira, M.R., Chao, L.W., Festa Neto, C., Silveira, P.S.P., Rivitti, E.A., Böhm, G.M., 2002. Web site for training nonmedical health-care workers to identify potentially malignant skin lesions and for teledermatology. Telemed. J. e-Health 8 (3), 323–332, <http://dx.doi.org/10.1089/15305620260353216>.

Park, E.K., Lim, J.Y., Yun, I.S., Kim, J.S., Woo, S.H., Kim, D.S., et al., 2016. Cranioplasty enhanced by three-dimensional printing: custom-made three-dimensional-printed titanium implants for skull defects. J. Craniofac. Surg. 27 (4), 943–949, <http://dx.doi.org/10.1097/SCS.0000000000002656>.

Pereira, C.A., Chao, L.W., Tavares, H., 2015. Alcohol abuse management in primary care: an e-Learning course. Telemed. J. E Health 21 (3), 200–206, <http://dx.doi.org/10.1089/tmj.2014.0042> Epub January 19, 2015.

Reddy, B., Salm, E., Bashir, R., 2016. Electrical chips for biological point-of-care detection. Annu. Rev. Biomed. Eng. 18, 329–355, <http://dx.doi.org/10.1146/annurev-bioeng-071813-104643>.

Retrieved July 11, 2016 from <https://www.youtube.com/watch?v=UAcCIsrAq70>.

Soirefmann, M., Comparin, C., Boza, J., Chao, L.W., Cestari, T.F., 2013. Impact of a cybertutor in dermatological teaching. Int. J. Dermatol. 52, 722–727, <http://dx.doi.org/10.1111/ijd.12030> Epub April 4, 2013.

Spitzer, M., 2015. The era of "digital colposcopy" will be here soon. J. Low Genit. Tract Dis. 19 (4), 273–274, <http://dx.doi.org/10.1097/LGT.0000000000000140>.

Sreelatha, O.K., Ramesh, S.V., 2016. Teleophthalmology: improving patient outcomes? Clin. Ophthalmol. 10, 285–295, <http://dx.doi.org/10.2147/OPTH.S80487>. eCollection 2016. Review.

Switz, N.A., D'Ambrosio, M.V., Fletcher, D.A., 2014. Low-cost mobile phone microscopy with a reversed mobile phone camera lens. PLoS One 9 (5), e95330, <http://dx.doi.org/10.1371/journal.pone.0095330. > eCollection 2014.

Taleb, A.C., Böhm, G.M., Avila, M., Chao, L.W., 2005. The efficacy of telemedicine for ophthalmology triage by a general practitioner. J. Telemed. Telecare 11 (S1), 83−85, <http://dx.doi.org/10.1258/1357633054461958>.

Topol, E., 2013. The Creative Destruction of Medicine. Basic Books, New York, NY.

Trindade, M.A.B., Wen, C.L., Festa, C.N., et al., 2008. Accuracy of store-and-forward diagnosis in leprosy. J. Telemed. Telecare 14 (2), 8−10, <http://dx.doi.org/10.1258/jtt.2008.071203>.

UC Berkeley showcases CellScope with 3D printed iPhone stand, June 5, 2012. Retrieved July 20, 2016 from <http://www.3ders.org/articles/20120605-uc-berkeley-showcases-cellscope-with-3d-printed-iphone-stand.html>.

Vieira Júnior, E.E., Chao, L.W., 2015. Training of beauty salon professionals in disease prevention using interactive tele-education. Telemed. e-Health 21, 55−61, <http://dx.doi.org/10.1089/tmj.2014.0044>. Epub Nove ember 17, 2014.

Will The Medical Tricorder From Star Trek Become Real?, May 25, 2016. Retrieved July 20, 2016 from <http://medicalfuturist.com/2016/05/25/will-the-medical-tricorder-from-star-trek-become-real/?ct=t(Newsletter_2014_07_177_17_2014)>.

Yang, F., Zhang, M., Bhandari, B., 2015. Recent development in 3D food printing. Crit. Rev. Food Sci. Nutr. Oct 19:0. [Epub ahead of print].

Ye, Y., Wang, J., Xie, Y., Zhong, J., Hu, Y., Chen, B., et al., 2014. Global teleophthalmology with iPhones for real-time slitlamp eye examination. Eye Contact Lens 40 (5), 297−300, <http://dx.doi.org/10.1097/ICL.0000000000000051>.

Your smartphone is millions of times more powerful that all of NASA's combined computing in 1969, October, 2015. Retrieved July 20, 2016 from <http://www.zmescience.com/research/technology/smartphone-power-compared-to-apollo-432/>.

Zeng, C., Xing, W., Wu, Z., Huang, H., Huang, W., 2016. A combination of three-dimensional printing and computer-assisted virtual surgical procedure for preoperative planning of acetabular fracture reduction. Injury 47, 2223−2227, pii: S0020-1383(16)30042-0. <http://dx.doi.org/10.1016/j.injury.2016.03.015>.

Zhang, Q., Ren, L., Shi, W., 2013. HONEY: a multimodality fall detection and telecare system. Telemed. J. E Health 19 (5), 415−429, <http://dx.doi.org?10.1089/tmj.2012.0109> Epub March 28, 2013.

Zhang, Y.S., Yue, K., Aleman, J., Mollazadeh-Moghaddam, K., Bakht, S.M., Yang, J., et al., 2016. 3D bioprinting for tissue and organ fabrication. Ann. Biomed. Eng. [Epub ahead of print]

# CHAPTER 9

# Analytics and Decision Support Systems in Global Health Informatics

**F. Martin-Sanchez and V. Aguiar-Pulido**
Weill Cornell Medicine, New York, NY, United States

## INTRODUCTION

Over the past ten years, the amount of data being generated in every possible field has increased in an unforeseen manner. This exponential growth is likely to be due to the proliferation of available methods to collect data, such as mobile devices, sensors, or online platforms. We have witnessed the rise of two new and interconnected disciplines related to data, i.e., big data and data science. However, in the health sector, relatively little is being done to exploit the full potential of these data, especially on what pertains to global health. In this chapter we review recent literature regarding the application of informatic methods focused on data analytics and decision support.

## The Scope of Public Health and Global Health

Even though there exists some controversy around the definition of the terms public health and global health, we hereby follow the perspective presented by Koplan et al. (2009). For these authors, global heath can be defined as "an area for study, research, and practice, which places a priority on improving health and achieving equity in health for all people worldwide". Therefore, it encompasses transnational health issues, social determinants of health, and it involves several disciplines and combines population-based prevention with individual-level clinical care. The US Institute of Medicine (IOM), on the other hand, describes the mission of public health as "fulfilling society's interest in assuring conditions in which people can be healthy" (Institute of Medicine, 1988). Hence, in this case, population health is emphasized as opposed to individual care. Fig. 9.1 depicts these differences in scope between the two disciplines.

## CONTENTS

195

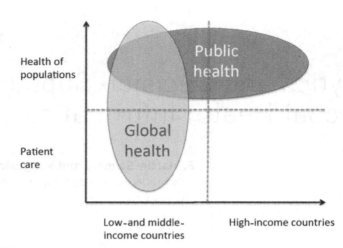

**FIGURE 9.1**
Global health versus public health—differences in scope.

Under these premises, it is important to bear in mind that global health issues should not be restricted to health concerns that affect more than one country; instead, global should be understood as any health problem involving multiple countries or those that are affected by transnational determinants. Some potential challenges include climate change, eradication of infectious diseases, tobacco control, obesity, and occupational health, among others; examples involving some of these challenges are introduced by Lester in his editorial (Lester, 2015). Furthermore, Mirza and colleagues propose a roadmap to set the grounds of informatics in global health (Mirza et al., 2012). Their main goal is to provide priorities for the efforts and highlight the current issues that are relevant for the field.

Following the proposed point of view, public health and global health can address domestic health problems, as well as cross-border issues, whereas international health addresses only the latter. Although global health usually refers to low- and middle-income countries, in this chapter, based on the previous definition, we include certain studies involving underrepresented populations (e.g., aborigine communities) within high-income countries.

## Enabling Technologies and Methods for Analytics and Decision Support in Global Health

To overcome the challenges mentioned in "The Scope of Public Health and Global Health" section, different technologies have arisen as potential enablers for global health interventions. In particular, the fast development of mobile technologies, including smartphones, tablets, and wearable

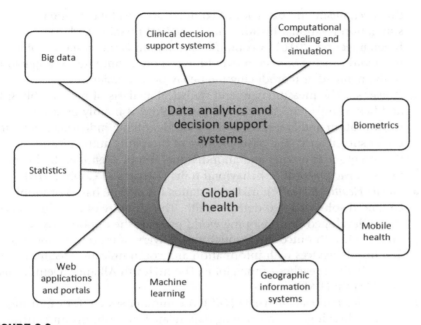

**FIGURE 9.2**
Data analytics and decision support systems in a global health context can be enabled by a series of technologies.

sensors, offers a promising avenue to bring the decision-making process closer to patients, engaging and empowering them. As shown in Fig. 9.2, many different technologies are relevant to data analytics and decision support systems within the context of global health. In low- and middle-income countries mobile technologies have penetrated especially well, being one of the key players in data collection and generation.

Subsequently, some basic definitions are provided for each of the technologies considered in Fig. 9.2.

- *Statistics*: The study of the collection, analysis, interpretation, presentation, and organization of data (Dodge, 2006).
- *Clinical Decision Support System* (*CDSS*): A health information technology application that provides clinicians, staff, patients, and other individuals with knowledge and person-specific information, intelligently filtered and presented at appropriate times, to enhance health and healthcare (Berner, 2009). Their main goal is to assist with clinical decision-making tasks.
- *Computational Modeling and Simulation*: Computational modeling is the use of mathematics, statistics, physics, and computer science to study

the mechanism and behavior of complex systems by computer simulation (National Institute of Biomedical Imaging and Bioengineering, 2016). A computer simulation is the imitation of a real-word process or system over time run on a computer or a network of them aimed at reproducing the behavior of a system.

- *Biometrics*: The measurement and statistical analysis of people's physical and behavioral characteristics. The technology is mainly used for identification and access control, or for identifying individuals who are under surveillance. The basic premise of biometric authentication is that everyone is unique and an individual can be identified by his or her intrinsic physical or behavioral traits (Rouse and Cobb, 2016).

- *Mobile Health (mHealth)*: mHealth stands for mobile-based or mobile-enhanced solutions that deliver health. The ubiquity of mobile devices in the developed or developing world presents the opportunity to improve health outcomes through the delivery of innovative medical and health services with information and communication technologies to the farthest reaches of the globe (The mHealth Alliance. Definitions of mHealth: HIMSS, 2016).

- *Geographic Information System (GIS)*: A computer system for capturing, storing, checking, and displaying data related to positions on Earth's surface. Geographical information systems can show many different kinds of data on one map. This enables people to see, analyze, and understand patterns and relationships more easily (Dunn, 2016).

- *Machine Learning (ML)*: Subfield of computer science that explores the study and construction of algorithms that can learn from and perform predictive analysis on data (Kohavi and Provost, 1998).

- *Web Application and Portals*: A web application is any application that uses a web browser as a client (i.e., the program the person uses to run the application) (Nations, 2016). A web portal (or simply portal) is a specially designed website that often serves as the single point of access for information. It can also be considered a library of personalized and categorized content. A portal helps in search navigation, personalization, notification, and information integration, and often provides features like task management, collaboration, and business intelligence and application integration (Techopedia. Web Portal, 2016).

- *Big Data*: The term Big Data is believed to have originated with Web search companies who had to query very large distributed aggregations of loosely structured data (O'Driscoll et al., 2013). This term has since been used to refer to the massive amounts of data collected over time, which are difficult to analyze and handle when using common database management tools (Dai et al., 2012). While the term may seem to reference the volume of data, this is not always the case. In a 2001 research report Gartner (formerly META Group) analyst

Doug Laney defined data growth challenges and opportunities as being three-dimensional, i.e., increasing volume (amount of data), velocity (speed of data in and out), and variety (range of data types and sources) (De Lissovoy, 2013). Much of the industry continues to use this "3 Vs" model for describing big data. In 2012 Gartner updated its definition as follows: "Big data is high volume, high velocity, and/or high variety information assets that require new forms of processing to enable enhanced decision making, insight discovery and process optimization." In addition, a new V for "Veracity" has been added by some organizations to describe it. The term Big Data may also refer to the technology (such as storage facilities, tools, and processes) that an organization requires for handling large amounts of data (Lusher et al., 2014). A more pragmatic definition describes Big Data in terms of a requirement for analytic applications to handle new types of data that (an organization) was not previously tracking.

Even though the term big data can mean different things to different groups of people, most of these would agree that global health may require massive, complex, distributed, and frequently dynamic sets of data (Martin-Sanchez and Verspoor, 2014). Hence, the dimension and complexity of these data will become challenging to analyze and thus create interesting opportunities for health organizations, especially from a computational perspective. These new sources of big data will need novel methods and approaches for their analysis and visualization, provided by the discipline of data analytics. To go from data to knowledge and make valuable use of the information, this needs to be incorporated in health organizations, e.g., integrating the discovered knowledge into decision support systems.

## Similarities and Differences Between Data Analytics and Decision Support

Both data analytics and decision support systems exploit the wealth of data that is increasingly becoming available for global health. However, these disciplines can be considered as sequential steps of a much bigger process. It may seem that there exists some overlap between these two phases, yet some differences are clearly noticeable.

Data analytics concerns mainly the discovery of hidden and meaningful patterns across data sources. The most frequently used methods involve statistical analysis and ML for predictive modeling purposes. Hence, it is grounded on artificial intelligence methods and data science techniques. On the other hand, decision support is a well-known discipline that has been in use since the 1970s. The main objective of decision support systems is to help humans in the decision-making process.

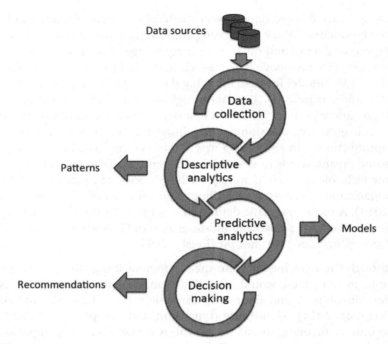

**FIGURE 9.3**
Workflow depicting the process of generating recommendations from data.

Fig. 9.3 shows a possible workflow from data collection to the suggestion of recommendations. Once data has been collected, data analytics is the first step toward turning data into knowledge. First, descriptive analytics is usually utilized to discover nontrivial trends and patterns. Second, predictive analytics is applied to generate computational models that represent the knowledge behind the data, which is relevant to address a particular health problem. Once we have that, we can create decision support systems that can help improve health by providing the final user with recommendations. This is especially important in low- and middle-income scenarios given their complexity in terms of how big the population is and the high number of diseases to be considered. In addition to this, because of counterfeit and the use of expired antibiotics, leading to antimicrobial resistance, we must face bigger challenges in comparison to first world countries.

In a similar manner to the "Model List of Essential Medicines" (World Health Organization, 2015) that was first launched by the World Health Organization in 1977, Dove and colleagues identify the need of creating an "Essential Diagnostics List" (Dove et al., 2015) that will take advantage of technologies such as those mentioned above. This list will allow establishing connections with health policies that cover multiple sectors, which may involve other sectors than just health (e.g., the use of omic technologies in the pharmaceutical field), as well as with science and ethics that are completely transparent and related to global society.

An overview of the application of informatics to global health is given in the work published by Dixon and collaborators (Dixon et al., 2015). In the rest of this chapter, we will focus on two specific fields, data analytics and decision support, presenting recent approaches that have been applied to global health informatics in relation to them.

## DATA ANALYTICS AND GLOBAL HEALTH

In this section we present some of the existing approaches in which data science is used to enhance global health. Starting from more traditional approaches based on statistics, toward big data approaches, and more complex techniques such as ML-based methods, we describe different examples of data analytics in regions belonging to the Middle East, Africa, and Asia. Fig. 9.4 depicts the main data analytics approaches reviewed subsequently.

### Statistical Analysis

Several researchers from Emory University carried out a study on the social representation of condoms (Winskell, 2011), which involved young citizens from six different regions of Africa between 10 and 24 years old. The ultimate goal of the study was to determine whether the increase of HIV/AIDS within young people was correlated with lack of information regarding condom use. This study presents a novel approach in the sense that it uses atypical data. Creative narratives submitted to a continent-wide script writing contest in 2005 that deal with HIV/AIDS were used as the qualitative data to be analyzed. The data from this study was stratified by sex, urban/rural location, and age. It is worth noting that people who participate in these contests, however, are likely to be better educated, more knowledgeable and

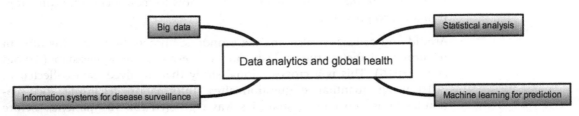

**FIGURE 9.4**

Overview of the main data analytics methods applied to global health.

motivated about HIV/AIDS than the average youth population, therefore the results obtained may present some biases. Nevertheless, since this will be the case for all the countries and it is known beforehand, conclusions will still be useful for the purpose of the study.

Two different approaches were followed in order to analyze the already processed data: a qualitative analysis, focused on theme-related analyses, and a narrative-based approach, which focused on the plot summary and keywords. In addition, quantifiable variables in relation to the narrative were also considered, such as, e.g., the sex of the characters involved. For the qualitative analysis, a series of descriptive codes were identified in the texts, such as condoms, abstinence, or religion, among others, based on a well-established methodology. For theme analysis, interpretive codes were identified. These codes were then investigated in relation to condom representation across the regions and by sex and age of author.

By analyzing this data, researchers have been able to identify misconceptions and misinformation regarding condom use, as well as some factors that could increase condom acceptability and strategies that could be useful regarding their promotion. This will contribute in understanding better the issues that these regions have to face regarding HIV/AIDS.

Another study that was conducted in Iran tried to estimate the economic burden that smoking-related major cancers involve (Rezaei et al., 2015). For this purpose, this study retrieved data from the GLOBOCAN databases and calculated a number of measures, such as potential life lost, cost of productivity lost, and mortality rate. This type of analysis, very popular for developed countries, is less common in developing countries. To determine which major cancers to include in the study, the authors reviewed the literature and took into consideration a World Health Organization report as well as those pinpointed by ICD-10. The population under study was selected within Iranian adults older than 35 years of age and was categorized into active smoker, former smoker, and never smoker. Results show how high the impact of smoking-attributable cancer deaths is. More specifically, the authors conclude that 2 out of 10 cancer deaths could be saved if there were no smokers in the country. Suggestions on how to reduce cigarette consumption are also provided.

Arnold and colleagues also analyze economic burden, however, this time in relation to diabetes, tuberculosis, and coprevalence in Kyrgyzstan (Arnold et al., 2016). This is a cross-sectional study that involved data collected in 2010 using quantitative questionnaires. Information regarding socioeconomic status and coping strategies was registered for 309 patients. Before analyzing the data, coarsened exact matching was applied to account for variation within groups of patients because of important differences in their socioeconomic background. This technique is nonparametric, allows imprecision, and tries to find groups of analogous covariates (with underlying distributions or using intuitive division) to perform matches within these groups even if they are not exact (as long as the observations can be coarsened into similar groups). For the analysis, the R language was used to obtain descriptive statistic measures. A logistic regression model was also built to assess the degree of association between the socioeconomic background and coping strategies. Gender, age, education, employment, and equivalent income in tertiles were used as independent variables in the models.

Results show that there are differences within the affected groups depending on the disease (tuberculosis, diabetes, or both) and also on the socioeconomic background. Based on these, the authors are able to conclude that out-of-pocket payments represent a clear burden for patients who suffer from diabetes or from both diabetes and tuberculosis. In addition to this, coming from a lower socioeconomic background poses an additional challenge in coping with the economic burden.

Finally, Musumari et al. present their work using traditional statistics on Myanmar workers who reside in Thailand (Musumari and Chamchan, 2016). This study tries to look at how this migration current affects HIV rates, since countries like Myanmar may not have so much information regarding HIV prevention, treatment, and care. Data from the PHAMIT-2 project was used as input to the analysis, which was collected in 2010 in the form of a structured questionnaire. The SPSS software was used to perform the statistical analyses. Descriptive statistic measures were first obtained and then bivariate analyses, such as Chi-square or Fisher's exact test (for categorical data), and Mann–Whitney $U$-test (for continuous data), were used to find associations between the covariates and the variable of interest. Three regression models that included a different number of variables were used to investigate HIV testing experience. To compare the results provided by these models, the Akaike Information Criterion was calculated.

The study revealed how little testing was being done for HIV among workers from Myanmar that moved to Thailand and helped identify potential barriers for this. In particular, there seems to be a trend for more educated citizens to use more healthcare services. Findings like these are essential to launch appropriate mechanisms for prevention, e.g., making HIV information accessible to those who have little or no formal education. Nevertheless, women seem to be more proactive than men, being twice as likely to have been tested for HIV. Thus, this suggests that information such as gender should also be taken into account when designing intervention programs. The authors conclude that safe sexual practices must be encouraged by educating citizens, ensuring that issues like language and confidentiality are addressed.

## ML for Prediction

More advanced models involved the use of ML techniques, such as artificial neural networks (Rau et al., 2016). Rau and colleagues retrieved data from the National Health Insurance Research Database (NHIRD) of Taiwan to apply data mining techniques with the goal of constructing a model that would allow predicting liver cancer after 6 years of being diagnosed of type II diabetes. Data from 2060 subjects were used to build artificial neural network and logistic regression prediction models using 10 variables (sex, age, alcoholic cirrhosis, nonalcoholic cirrhosis, alcoholic hepatitis, viral hepatitis, other types of chronic hepatitis, alcoholic fatty liver disease, other types of fatty liver disease, and hyperlipidemia). The authors used the software STATISTICA to obtain the models for liver cancer risk prediction, including in these only those factors that were reported statistically significant after performing Chi-square in a sequential fashion. On the basis of the followed approach, the authors obtained three types of submodels that were used for diverse clinical purposes.

Results show that the artificial neural network models are superior in performance and, hence, the authors conclude that the proposed approach could be used to predict liver cancer in patients with diabetes. More specifically, of all the variables initially considered, only alcoholic cirrhosis, cirrhosis, viral hepatitis, chronic hepatitis, and hyperlipidemia were included in the final models. The resulting approach was implemented and made available through a user-friendly web interface. Finally, when discussed with clinicians, the model obtained in this work seemed to be of assistance to them to advise potential liver cancer patients.

## Information Systems for Disease Surveillance

Other interesting approaches regarding data analytics in Global Health may involve the development of information systems for disease surveillance and antimicrobial resistance. In this sense, Ohrt et al. reviewed the existing literature to identify key components that should be part of these systems, and presented a list of information systems that were developed to support surveillance for malaria elimination (Ohrt et al., 2015). To identify the desired characteristics of this type of system, the authors interviewed 21 subjects from different areas of expertise, such as malaria field experts, surveillance specialists, experts in geographic information systems, information technology experts and experts in the control and eradication of malaria and other diseases. The features considered as important by Ohrt et al. include rapid and complete reporting of data, incorporation of additional data from external resources to the malaria program and health system, accessible data storage and management, combination of automated and expert analysis, customized output and feedback, and targeted response. In addition, existing surveillance systems applied in different countries (Cambodia, China, Solomon Islands/Vanuatu, Swaziland, Thailand, Zambia, Tanzania) were described in this review. The authors conclude giving a set of recommendations after detecting several important issues that require attention.

With the aim of studying influenza antiviral resistance, Dafilis and collaborators developed a model that takes into account "real-world" constraints when modeling potential scenarios (Dafilis et al., 2012). More specifically, the authors follow a Latin hypercube sampling approach to investigate possible ranges of parameters related to wild type and resistant viruses, in addition to intervention efficacy, target coverage, and distribution capacity. The proposed model is based on the classic deterministic susceptible-exposed-infectious-recovered paradigm, modified by adding a component to take into account the rise of a drug-resistant strain. This paradigm assumes that any individual is susceptible at the epidemic outbreak and, therefore, vulnerable to infection if exposed to an infectious case. Once the individual recovers from the infection, he/she is assumed to be resistant to reinfection.

Models like the one proposed by Dafilis et al. could be useful to guide surveillance activities regarding preparedness and response and, thus, leading to the improvement of policies. In addition, these authors claim that they were able to obtain fundamental drivers and signals of resistance emergence, which may improve the collection of information during a pandemic response. Some conclusions extracted from this study highlight the need of a globally coordinated initiative regarding active surveillance of transmissible drug-resistant influenza mutants.

## Big Data

In general, we can observe there is a clear tendency to generate increasing volumes of very large data in the clinical field. This is what is being currently called big data in healthcare. However, according to Wyber et al., little progress has been done in data analytics although following appropriate guidelines could help improve the quality, quantity, storage, and analysis of this type of data (Wyber et al., 2015). Therefore significant improvements could be achieved in many health outcomes. Health-related big data could aid in the development of learning systems and in achieving a more precise management of the individuals, leading to improving the health of the overall population. Especially in low-and middle-income countries, developing approaches that involve data analytics in big data could represent a powerful resource. There are, nonetheless, several challenges to face given that most health records are still stored in paper and that these countries are more prone to fragmentation and misuse.

**Table 9.1** Overview of Data Analytics Approaches Classified by Method and Application

| | Methods | | | |
|---|---|---|---|---|
| **Applications** | **Statistical Analysis** | **Machine Learning** | **Information Systems and Modeling** | **Big Data** |
| Health literacy | Winskell (2011) | | | |
| Economic burden of disease | Rezaei (2015) | | | |
| | Arnold et al. (2016) | | | |
| Health effects of migration | Musumari (2016) | | | |
| Disease prediction | | Rau (2016) | | |
| Disease surveillance and antimicrobial resistance | | | Ohrt (2015) | |
| | | | Dafilis (2012) | |
| Health statistics | | | | Wyber (2015) |

An interesting and promising initiative to tackle this problem has been started in India. Since 2010, the government has been providing citizens with cards and unique numbers associated to biometric identification, allowing for health and social data generation and monitoring. This system, called Aadhaar, could represent a milestone toward obtaining a more reliable and extensive set of health statistics. Finally, this may help to significantly improve the planning and delivery of public health interventions.

Gathering all this data and identifying potential predictors of disease is only the first step in the application of data science techniques to global health. Subsequently, different decision support systems applied to global health will be described.

## Summary of Data Analytics Applications in Global Health

In this section, we have summarized all the approaches described above in a table, classifying them not only by the type of method used, but also taking into account how they were applied (see Table 9.1).

Next, different examples of how the results obtained using data analytics can be integrated into the decision-making process in global health informatics are described.

## DECISION SUPPORT SYSTEMS AND GLOBAL HEALTH

In this section we present some of the existing approaches that concern the use of decision support systems. Starting from patient-centered approaches,

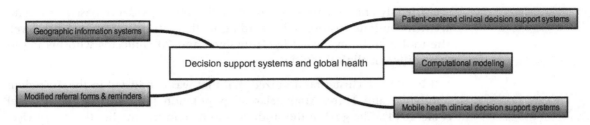

**FIGURE 9.5**
Overview of the main decision support methods applied to global health.

toward provider-centered ones, we describe different examples of CDSSs that
have been tested on low- and middle-income countries, as well as underrepresented populations from high-income countries. Fig. 9.5 depicts the main
decision support methodologies reviewed below.

## Patient-Centered and Web-Based CDSSs

Subsequently, patient-centered CDSSs will be presented. In general, these
tools are evidence-based and were conceived with the aim of enabling
patients to take part in the decision-making process regarding their health.
By showing the patient all the alternatives and with the clinician's guidance,
these systems enhance the overall decision-making process.

Following this philosophy, a group of researchers from Australia in collaboration with researchers from the United Kingdom developed a web-based
approach to facilitate the process of making decisions by providing personalized
risk assessment (Collins et al., 2014). Hence, healthcare professionals are
offered an easy-to-use tool that will help them assess a woman's risk to breast
cancer; additionally, the tool will provide appropriate information regarding
the person's condition. To build the system, these researchers examined qualita-
tively how risk assessment and management of breast cancer was being currently carried out. As a result, some key factors, challenges, and limitations were
identified, setting the ground for the development of this kind of tool and
increasing the women's degree of understanding of their own risks and options.

A more recent approach is known as IPrevent (Collins et al., 2016). This
tool, also designed for managing risk in interventions, combines two different well-validated models (IBIS and BOADICEA) to provide evidence-based,
risk-adapted suggestions for risk management. The two risk models on which
the decision support system relies take into account family history at different degrees of relatedness and include different cancer types, as well as other
factors such as body mass index, reproductive factors for IBIS, or cancer
pathology characteristics for BOADICEA. In addition, IPrevent provides a

user-friendly and very intuitive interface, including evidence-based risk management options customized according to the patient's estimated risk level. The tool is to be evaluated by 70 women and 20 clinicians who belong to three different clinical settings.

TORPEDO is a cluster-randomized trial that was conducted in the Australian Aboriginal and Torres Strait Islander population (Peiris et al., 2015; Patel et al., 2014). The goal of the study was to determine whether the use of electronic decision support systems improved the risk factor measurements and the medications indicated by the guidelines for patients at high cardiovascular risk. This study involved more than 38,000 people and 60 health services and decision support systems were used in combination with audit and feedback strategies. Monthly, peer-ranked performance feedback was also collected through a web portal. Although there is a 10% improvement regarding the screening for cardiovascular disease risk and there are relevant results in terms of treatment escalation, there does not seem to be any significant difference involving the prescription of recommended medicines. Therefore, this appears to indicate that computerized tools may be useful, especially concerning preventive treatments.

Another cluster-randomized trial related to cardiovascular disease was carried out by Ajay and colleagues (Ajay et al., 2014). This trial involved people from 40 villages from two different regions of rural China and India. In this case an electronic decision support system, based on a mobile device, will be developed to help community health workers with their high-risk patients. This system will be built as a pilot and will be based on the intervention strategy applied during the year the study lasts.

## Computational Modeling

Construct is a tool that allows simulating community health centers based on data that was gathered at an earlier stage (Carney et al., 2015). It features a multiagent model of "social" network evolution and allows building virtual scenarios to study specific behaviors and generate hypotheses involving organizational dynamics. More specifically, it is based on dynamic network theory and estimates interactions (such as gain, loss, spread, or retention of knowledge within an organization) by means of probabilistic models. The objective of the research conducted by its developers was to investigate whether there existed simple, nonlinear processes that could be attributable to team or group behavior using computational approaches. The proposed approach was tested on a healthcare facility using information from cancer screening, suggesting that a healthcare organization can be viewed as a complex adaptive entity. These findings may have important implications regarding knowledge acquisition, retention, and sharing in CDSSs.

The Strategic Multi-Attribute Ranking Tool for Vaccines (SMART Vaccines) (Madhavan et al., 2013) builds on the ranking that was obtained and tested for streptococcal infections by the IOM in 2012. This multiattribute utility model was further refined to obtain an enhanced version of the software that was publicly released. The main goal of this tool is to provide a basis for discussion regarding the prioritized list suggested by the system.

## Mobile Health CDSSs

Mobile technology is increasingly becoming key in rural healthcare environments and several studies show its potential specially when used by frontline health providers (Agarwal et al., 2015). Taking advantage of this technology, the developers of SMARTHealth build a CDSS to facilitate Systematic Appraisal Referral and Treatment of cardiovascular disease risk for the rural parts of India (Praveen et al., 2013). This system includes information from Indian and international guidelines, the World Health Organization and the International Society of Hypertension, Indian National Program for Prevention and Control of Cancer, and Diabetes Cardiovascular and Stroke guidelines. The intervention will be carried out as a cluster-randomized controlled trial involving 18 primary health centers from 54 villages and 15,000 adults older than 40 years of age and at high cardiovascular risk. More details about the system's evaluation are provided in a later publication by the same authors (Praveen et al., 2014), who conclude that mobile health approaches are powerful and that they can be successfully integrated in the current Indian primary healthcare system. Furthermore, the proposed tablet-based decision support system may be effective to enhance cardiovascular disease detection, prevention, and management in the setting being studied. An updated version of the tool was published by Raghu et al. (2015).

The Academic Model Providing Access to Healthcare (AMPATH) was developed in collaboration among several institutions from Kenya and North America and applied to western Kenya (Vedanthan et al., 2015). Within this initiative, mobile health has been used in a rural environment to enhance the healthcare system by providing decision support and record-keeping functions. This type of technology is supposed to engage patients, increasing the rate of retention and adherence to care. The model was tested on patients with cardiovascular disease; however, more research is required to accurately evaluate the impact of these technologies in rural settings. Nonetheless, the authors believe that involving the complete care cascade is the first step toward secondary prevention of cardiovascular disease in low-income countries.

Another example of an application of mobile health to a rural environment was presented by Velez and collaborators (Velez et al., 2014). These authors developed an eHealth delivery platform named mClinic, which collects data to be used for patient care management, program evaluation and monitoring, and decision-making. Given that midwives in rural Ghana are the ones who lead the healthcare system, this platform targeted them specifically. mClinic was evaluated by means of a usability questionnaire and interviews with the final users, which allowed identifying relevant issues that must be improved.

MASK-rhinitis follows a sentinel network-based approach for syndrome surveillance in allergic rhinitis (Bousquet et al., 2015). This system consists of a CDSS, which is fed with information from the ARIA 2015 guideline for allergic rhinitis and asthma comorbidity. It was developed under a European initiative for Active and Healthy Aging and it uses three main strategies for electronic monitoring of allergic diseases: assessing disease control daily using a visual analogue scale that is cell phone—based, the Control of Allergic Rhinitis and Asthma Test (CARAT), and an allergy screening that uses online tools. The authors conclude that approaches that involve these strategies in combination with the use of CDSSs will help improve policies and standards.

## Modified Referral Forms and Reminders

Jenkins and colleagues conducted a review on the effectiveness of interventions conceived to decrease the use of imaging for low-back pain (Jenkins et al., 2015). In their research, they found that, in a hospital environment, utilizing CDSSs that involved a modified referral form or including targeted reminders to general practitioners with suitable guidelines helped reducing the amount of imaging tests requested. Applying this on a wider scale would result in significantly decreasing medical costs and, therefore, could be considered as potentially cheaper interventions.

Green and colleagues elaborated a research protocol for a study aimed at evaluating the impact of deploying a reminder system for healthcare providers on the use of a specific therapy for HIV patients in western Kenya (Green et al., 2015). The reminder systems considered in this study must be part of CDSSs. More specifically, the role of the clinical support system used will be to support the healthcare provider's performance by providing them with a sheet that summarizes essential details and offers tailored care suggestions, thus allowing a less trained provider to offer higher-quality standardized care. The main goal of the study is to determine whether the use of these systems ameliorated the proportion of preventive treatment in patients with both tuberculosis and HIV comorbidity who lived in a tuberculosis endemic region. Twenty public medical centers will participate in this cluster-randomized study.

## Geographic Information Systems

Noor et al. developed a spatial decision support framework aimed at seasonal malaria chemoprevention for Sahelian countries that belong to the Nouakchott initiative (Noor et al., 2015). This framework integrates information from two different malaria risk maps with information regarding seasonality, population structure based on age, urbanization, and malaria endemicity. Based on the results obtained after analyzing nearly 600 Sahelian health districts, the authors conclude that the framework is likely to be beneficial for the inhabitants of this region; however, more resources will be necessary in order to scale up the coverage to other countries that could potentially benefit from the proposed approach.

## Summary of Decision Support Applications in Global Health

In this section, we have summarized in a table all the approaches previously described, classifying them not only by the type of method used, but also taking into account how they were applied (see Table 9.2).

Subsequently, conclusions regarding the overall chapter will be drawn and future possible directions will be proposed.

## CONCLUSIONS AND FUTURE DIRECTIONS

After reviewing the recent literature on the application of informatics in Global Health, particularly through the lenses of data analytics and decision-making, it could be worth discussing whether there are some common features to the projects hereby reported. If so, these could serve as a compendium of good practices, which could guide the development and implementation of future projects.

In terms of common characteristics and requirements for Global Health Informatics, in the area of decision support, we can point out that socio-politic-economic realities of low- and middle-income countries present important challenges, such as the level of poverty in the population, the very low national spending on health, the impact derived from citizens incurring in significant out-of-pocket health expenses, a limited outpatient health insurance and poor health infrastructures (human resources, networks, medication stock, facilities, equipment).

Particularities derived from this reality suggest that if we are to succeed in implementing Global Health Informatics programs in these countries or even in underserved, isolated (rural, remote) regions of developed countries we need to pay more attention and conveniently address a number of requirements. These include the need for systems to be very simple and easy to use,

**Table 9.2** Overview of Decision Support Approaches Classified by Method and Application

| Applications | Methods | | | | |
| --- | --- | --- | --- | --- | --- |
| | Patient-Centered Clinical Decision Support Systems—Web Based | Computational Modeling | Mobile Health Clinical Decision Support Systems | Modified Referral Forms and Reminders | Geographic Information Systems |
| Personalized medicine, risk assessment, and prevention | Collins (2014) Collins (2016) Patel et al. (2014) and Peiris et al. (2015) | | Praveen (2013, 2014) | | |
| Screening | | Carney et al. (2015) | | | |
| Treatment and continuous disease monitoring | Ajay et al. (2014) | | Vedanthan (2015) Velez (2014) Agarwal (2015) | Green et al. (2015) | |
| Effectiveness of interventions | | | | Jenkins (2015) | |
| Research—Vaccine development prioritization | | Madhavan et al., (2013) | | | |
| Syndrome surveillance and sentinel networks | | | Bousquet et al. (2015) | | |
| Chemoprevention in infectious diseases | | | | | Noor (2015) |

as the general level of technical and health literacy in these populations is relatively low. Also, there is a clear need to pay attention and conform to the realities of clinical practice. A strong evidence base (supported by professional bodies) is required in the knowledge that supports all the reasoning and processing of decision support systems. Technologies of choice have to be reliable, for instance, in terms of connectivity and battery use. Language can be a barrier and in these places, very frequently, many local dialects and languages are spoken. This needs to be taken into account when developing user interfaces, preparing documentation, and providing training. Along these lines, many cultural issues can pose challenges for the smooth operation of Information Technology systems, including traditions, food constraints, or spiritual aspects. There is always a resistance to change that needs to be overcome, requiring special consideration to properly embed the new systems in routine professional practice and workflows. Finally, a strong support and commitment from managers and high-level administrative officers represents a critical requirement to complete implementation within timeframe and budget.

Although global health informatics is a relatively new discipline, it has helped to achieve important milestones. By using technology (e.g., access to health services by population that come from a low- and middle-income), socioeconomic background has been improved. In particular, aspects related to the massive use of data (known as big data) and mobile technologies offer enormous potential to improve the analytical and support decision-making processes. Bearing this in mind, different possible future directions may be followed.

With the rise of diseases like Zika, which have devastating consequences, a high rate of contagiousness and become very fast a widespread and global problem, it is clear that more local surveillance programs are required. These programs need to be not only robust, but must also exhibit a faster response by reducing the timeframe required to identify new cases. In addition a greater level of engagement and training of healthcare professionals will be required. In this context, CDSSs such as those described in this chapter will be crucial to advance in the implementation of diagnosis and control measurements. Programs involving the training and education of the population in general will be essential for disease prevention.

Data analytics and decision support can become very relevant in the field of global mental health. Retrieving, processing, and analyzing information from multiple heterogeneous sources (including electronic health records) and the creation of decision support systems based on the knowledge discovered from it may have a great potential to improve this field; especially concerning the detection, modeling and prediction of suicide risk, depression or conditions like epilepsy.

Another issue that becomes of increasing importance is the lack of equity. This must be fought against by providing access to reliable knowledge regarding health problems and care. Not only low- and middle-income populations have to be accounted for but also underserved populations in high-income countries should be considered. In this sense, resources such as telemedicine or mobile technologies can be powerful. Hence, these technologies must be further explored, developed, and evaluated with the aim of connecting health providers, citizens, and information resources.

Mobile health, which is an example of participatory technology and is becoming increasingly popular, as we have seen along the chapter, may be used for tasks such as postsurgical monitoring or chronic disease management. This type of approach can reduce the cost of health services and yet improve its quality. Following the same train of thought, mobile applications and wearable sensors, which are part of a greater field known as the "quantified-self" (Almalki et al., 2015), may prove to be a potent resource for real-time monitoring of people; in particular for assessing exposure to environmental risk factors at the individual level, measuring factors such as physical activity, sleep quality, or mood (Martin-Sanchez et al., 2014).

Although commercial electronic medical record platforms are still very expensive, both to purchase and maintain, there exist a number of open software solutions that can be adopted by low- and middle-income countries to improve how clinical information is managed and exchanged. Some examples of software that are free and widely available are openMRS—an electronic health record, and openHIE for health information exchange (Center for Biomedical Informatics, 2016). A wider use of these platforms will enable a more personalized approach to care as well as improvements in the secondary use of clinical data for research and global health purposes.

To overcome existing limitations regarding the access of populations and communities to healthcare and prevention, technology is becoming a key player. To achieve these goals, there is an urgent need for training the next generation of global health informaticians. This training will involve ensuring that global health informaticians have competence in effective project management (e.g., incorporating clinical information from early stages in the process), publishing research results (to increase the evidence base), and in applying implementation science including change management theories.

# References

Agarwal, S., Perry, H.B., Long, L.A., Labrique, A.B., 2015. Evidence on feasibility and effective use of mHealth strategies by frontline health workers in developing countries: systematic review. Trop. Med. Int. Health 20 (8), 1003–1014.

Ajay, V.S., Tian, M., Chen, H., Wu, Y., Li, X., Dunzhu, D., et al., 2014. A cluster-randomized controlled trial to evaluate the effects of a simplified cardiovascular management program in Tibet, China and Haryana, India: study design and rationale. BMC Public Health 14, 924.

Almalki, M., Gray, K., Sanchez, F.M., 2015. The use of self-quantification systems for personal health information: big data management activities and prospects. Health Inf. Sci. Syst. 3 (Suppl. 1), S1, HISA Big Data in Biomedicine and Healthcare 2013 Con.

Arnold, M., Beran, D., Haghparast-Bidgoli, H., Batura, N., Akkazieva, B., Abdraimova, A., et al., 2016. Coping with the economic burden of diabetes, TB and co-prevalence: evidence from Bishkek, Kyrgyzstan. BMC Health Serv. Res 16, 118.

Berner, E.S., 2009. Clinical Decision Support Systems: State of the Art. Agency for Healthcare and Quality, Rockville, MD.

Bousquet, J., Schunemann, H.J., Fonseca, J., Samolinski, B., Bachert, C., Canonica, G.W., et al., 2015. MACVIA-ARIA Sentinel NetworK for allergic rhinitis (MASK-rhinitis): the new generation guideline implementation. Allergy 70 (11), 1372−1392.

Carney, T.J., Morgan, G.P., Jones, J., McDaniel, A.M., Weaver, M.T., Weiner, B., et al., 2015. Hypothesis generation using network structures on community health center cancer-screening performance. J. Biomed. Inform. 57, 288−307.

Center for Biomedical Informatics, 2016. Global Health Initiatives 2016. Available from: <https://www.regenstrief.org/cbmi/areas-excellence/global-health/initiatives/> (cited August 1, 2016).

Collins, I.M., Bickerstaffe, A., Ranaweera, T., Maddumarachchi, S., Keogh, L., Emery, J., et al., 2016. iPrevent(R): a tailored, web-based, decision support tool for breast cancer risk assessment and management. Breast Cancer Res. Treat. 156 (1), 171−182.

Collins, I.M., Steel, E., Mann, G.B., Emery, J.D., Bickerstaffe, A., Trainer, A., et al., 2014. Assessing and managing breast cancer risk: clinicians' current practice and future needs. Breast 23 (5), 644−650.

Dafilis, M.P., Moss, R., McVernon, J., McCaw, J., 2012. Drivers and consequences of influenza antiviral resistant-strain emergence in a capacity-constrained pandemic response. Epidemics 4 (4), 219−226.

Dai, L., Gao, X., Guo, Y., Xiao, J., Zhang, Z., 2012. Bioinformatics clouds for big data manipulation. Biol. Direct 7, 43, discussion.

De Lissovoy, G., 2013. Big data meets the electronic medical record: a commentary on "identifying patients at increased risk for unplanned readmission". Med. Care 51 (9), 759−760.

Dixon, B.E., Pina, J., Kharrazi, H., Gharghabi, F., Richards, J., 2015. What's past is prologue: a scoping review of recent public health and global health informatics literature. Online J. Public Health Inform. 7 (2), e216.

Dodge, Y., 2006. The Oxford Dictionary of Statistical Terms. Oxford University Press, Oxford.

Dove, E.S., Barlas, I.O., Birch, K., Boehme, C., Borda-Rodriguez, A., Byne, W.M., et al., 2015. An appeal to the global health community for a tripartite innovation: an "essential diagnostics list," "health in all policies," and "see-through 21(st) century science and ethics". OMICS 19 (8), 435−442.

Dunn, M.G., 2016. Definition of geographic information system: National Geographic Society. Available from: <http://nationalgeographic.org/encyclopedia/geographic-information-system-gis/> (cited August 1, 2016).

Green, E.P., Catalani, C., Diero, L., Carter, E.J., Gardner, A., Ndwiga, C., et al., 2015. Do clinical decision-support reminders for medical providers improve isoniazid preventative therapy prescription rates among HIV-positive adults? Study protocol for a randomized controlled trial. Trials 16, 141.

Institute of Medicine, 1988. The Future of Public Health. National Academy Press, Washington, DC.

Jenkins, H.J., Hancock, M.J., French, S.D., Maher, C.G., Engel, R.M., Magnussen, J.S., 2015. Effectiveness of interventions designed to reduce the use of imaging for low-back pain: a systematic review. Can. Med. Assoc. J. 187 (6), 401–408.

Kohavi, R., Provost, F., 1998. Glossary of terms. Mach. Learn. 30, 271–274.

Koplan, J.P., Bond, T.C., Merson, M.H., Reddy, K.S., Rodriguez, M.H., Sewankambo, N.K., et al., 2009. Towards a common definition of global health. Lancet 373 (9679), 1993–1995.

Lester, R.T., 2015. Connecting patient care to global health trends by health app analytics. Public Health Action 5 (4), 203.

Lusher, S.J., McGuire, R., Van Schaik, R.C., Nicholson, C.D., De Vlieg, J., 2014. Data-driven medicinal chemistry in the era of big data. Drug Discov. Today 19 (7), 859–868.

Madhavan, G., Sangha, K., Phelps, C., Fryback, D., Rappuoli, R., Martinez, R.M., 2013, Committee on Identifying and Prioritizing New Preventive Vaccines for Development, Phase II, Board on Population Health and Public Health Practice, Board on Global Health, Institute of Medicine, et al., 2013. Ranking Vaccines: A Prioritization Software Tool: Phase II: Prototype of a Decision-Support System. The National Academies Press, Washington, D.C.

Martin Sanchez, F., Gray, K., Bellazzi, R., Lopez-Campos, G., 2014. Exposome informatics: considerations for the design of future biomedical research information systems. J. Am. Med. Inform. Assoc 21 (3), 386–390.

Martin-Sanchez, F., Verspoor, K., 2014. Big data in medicine is driving big changes. Yearb. Med. Inform. 9, 14–20.

Mirza, M., Kratz, M., Medeiros, D., Pina, J., Richards, J., Zhang, X., et al., 2012. Building the foundations of an informatics agenda for global health-2011 workshop report. Online J. Public Health Inform. 4 (1).

Musumari, P.M., Chamchan, C., 2016. Correlates of HIV testing experience among migrant workers from Myanmar residing in Thailand: a secondary data analysis. PLoS One 11 (5), e0154669.

National Institute of Biomedical Imaging and Bioengineering, 2016. Glossary of science education: NIH. Available from: <https://www.nibib.nih.gov/science-education/glossary-g-42826> (cited August 1, 2016).

Nations, D., 2016. What is a web application? Available from: <http://webtrends.about.com/od/webapplications/a/web_application.htm> (cited August 1, 2016).

Noor, A.M., Kibuchi, E., Mitto, B., Coulibaly, D., Doumbo, O.K., Snow, R.W., 2015. Sub-national targeting of seasonal malaria chemoprevention in the Sahelian countries of the Nouakchott initiative. PLoS One 10 (8), e0136919.

O'Driscoll, A., Daugelaite, J., Sleator, R.D., 2013. 'Big data', Hadoop and cloud computing in genomics. J. Biomed. Inform. 46 (5), 774–781.

Ohrt, C., Roberts, K.W., Sturrock, H.J., Wegbreit, J., Lee, B.Y., Gosling, R.D., 2015. Information systems to support surveillance for malaria elimination. Am. J. Trop. Med. Hyg. 93 (1), 145–152.

Patel, B., Patel, A., Jan, S., Usherwood, T., Harris, M., Panaretto, K., et al., 2014. A multifaceted quality improvement intervention for CVD risk management in Australian primary healthcare: a protocol for a process evaluation. Implement. Sci. 9, 187.

Peiris, D., Usherwood, T., Panaretto, K., Harris, M., Hunt, J., Redfern, J., et al., 2015. Effect of a computer-guided, quality improvement program for cardiovascular disease risk management in primary health care: the treatment of cardiovascular risk using electronic decision support cluster-randomized trial. Circ. Cardiovasc. Qual. Outcomes 8 (1), 87–95.

Praveen, D., Patel, A., McMahon, S., Prabhakaran, D., Clifford, G.D., Maulik, P.K., et al., 2013. A multifaceted strategy using mobile technology to assist rural primary healthcare doctors and frontline health workers in cardiovascular disease risk management: protocol for the SMARTHealth India cluster randomised controlled trial. Implement. Sci. 8, 137.

Praveen, D., Patel, A., Raghu, A., Clifford, G.D., Maulik, P.K., Mohammad Abdul, A., et al., 2014. SMARTHealth India: development and field evaluation of a mobile clinical decision support system for cardiovascular diseases in rural India. JMIR Mhealth Uhealth 2 (4), e54.

Raghu, A., Praveen, D., Peiris, D., Tarassenko, L., Clifford, G., 2015. Engineering a mobile health tool for resource-poor settings to assess and manage cardiovascular disease risk: SMARThealth study. BMC Med. Inform. Decis. Mak. 15, 36.

Rau, H.H., Hsu, C.Y., Lin, Y.A., Atique, S., Fuad, A., Wei, L.M., et al., 2016. Development of a web-based liver cancer prediction model for type II diabetes patients by using an artificial neural network. Comput. Methods Programs Biomed. 125, 58−65.

Rezaei, S., Akbari Sari, A., Arab, M., Majdzadeh, R., Mohammadpoorasl, A., 2015. Estimating economic burden of cancer deaths attributable to smoking in Iran. J. Res. Health Sci. 15 (4), 228−233.

Rouse, M., Cobb, M., 2016. Definition of biometrics: TechTarget. Available from: <http://search-security.techtarget.com/definition/biometrics> (cited August 1, 2016).

Techopedia. Web Portal, 2016. Available from: <https://www.techopedia.com/definition/17352/web-portal> [cited August 1, 2016).

The mHealth Alliance. Definitions of mHealth: HIMSS, 2016. Available from: <http://www.himss.org/definitions-mhealth> (cited August 1, 2016).

Vedanthan, R., Kamano, J.H., Bloomfield, G.S., Manji, I., Pastakia, S., Kimaiyo, S.N., 2015. Engaging the entire care cascade in Western Kenya: a model to achieve the cardiovascular disease secondary prevention roadmap goals. Glob. Heart 10 (4), 313−317.

Velez, O., Okyere, P.B., Kanter, A.S., Bakken, S., 2014. A usability study of a mobile health application for rural Ghanaian midwives. J Midwifery Wom. Health 59 (2), 184−191.

Winskell, K., Obyerodhyambo, O., Stephenson, R., 2011. Making sense of condoms: social representations in young people's HIV-related narratives from six African countries. Soc. Sci. Med. 72 (6), 953−961.

World Health Organization, 2015. WHO model lists of essential medicines. Available from: <http://www.who.int/medicines/publications/essentialmedicines/en/> (cited August 1, 2016).

Wyber, R., Vaillancourt, S., Perry, W., Mannava, P., Folaranmi, T., Celi, L.A., 2015. Big data in global health: improving health in low-and middle-income countries. Bull. World Health Organ. 93 (3), 203−208.

# CHAPTER 10

# Patient Engagement and Digital Health Communities

**H. de Fátima Marin[1,2] and C. Delaney[3,4,5]**

[1]Federal University of São Paulo, São Paulo, Brazil
[2]Director, Innovation & IT, Hospital Sirio Libanês, São Paulo, Brazil
[3]American College of Medical Informatics, Bethedsa, MD, United States
[4]University of Minnesota, Minneapolis, MN, United States
[5]American Academy of Nursing, Washington, DC, United States

## CONTENTS

## PATIENT ENGAGEMENT

Patient engagement is being used to describe a range of possibilities and strategies to involve patients and citizens with the purpose to obtain their participation and collaboration with the healthcare team in managing their own health and wellness.

Several factors contribute to the deployment of patient engagement strategies such as the need for shared decision-making with citizen and patients' wish to deal with their own wellness, the explosive growth of information and knowledge available on the Internet for citizens and patients, and the realities of the shortages of healthcare workforce particularly in comparison to the increasing demand for healthcare and the financial constraints faced by countries and healthcare provider institutions.

Once again, we are testimonies to how the advent of the Internet is changing our way of living, working, and making decisions in daily activities. Engagement of patients involves not only their partnership but also patients leading and determining healthcare needs and interventions, in many cases having patient input spanning research as well as medical and health care. For example, in the United States, The Patient-Centered Outcomes Research Institute's (PCORI)'s strong emphasis on engaging patients and the broader healthcare community in all their work is evident in the criteria used to determine what research is funded. PCORI provides research funding that engages patients and other stakeholders (http://www.pcori.org/).

**Global Health Informatics.**

Engagement ranges from patients determining relevant effectiveness research priorities to full engagement in the research and dissemination of findings.

Thus the current opportunities are to align and engage patients with the challenges initiated by advances in science and technology in global healthcare, and demand innovative solutions to address complex problems that are common across countries such as the aging of the population, chronic diseases, high costs of care delivery, economic constraints to assuring continuity, quality, and cost benefits of treatments and care, and emerging and reemerging diseases.

Consequently, health systems around the world face considerable challenges in providing healthcare services. The healthcare systems in the world are experiencing significant crises and considerable challenges (Neto, 2014).

Factors contributing to this situation of crises include: aging of the population and the consequent change in the way that health services are requested and delivered:

- Changes of the epidemiological profile of populations (decrease of infectious diseases with higher prevalence on noncommunicable diseases, resurgence of infectious diseases of the past such as tuberculosis, dengue, cholera, and the emergence of new infectious diseases such as AIDS, Ebola, Zika, and hantaviruses);
- Knowledge diffusion generating important modifications in the willingness of people to engage in consumer actions and health services;
- Medicalization trend that encourages people as consumers of medical services and actions to pursue health status;
- Corporatization of health professionals to work by division and tasks that deconstruct the necessary inter and transdisciplinarity in health and can result in turning the patient into a *victim* of multiple professionals;
- Technological revolution (not ICT—Information and Communication Technology) that offers new alternatives without necessarily abandoning the previous one, causing extra costs; and
- Pursuit of equity, where everyone has access to everything, and increased requests for state financial coverage.

Among all factors, aging of the population can be highlighted as an example of how each of these factors affects the whole healthcare system. Due to progressive aging and the changing expectations of people to receive all that is available in healthcare to extend life, people's greatest expenditure on health occurs during the final decade of their lives. For this reason, health systems allocate a significant part of their resources to the provision of curative and palliative services (Fernandez and Oviedo, 2011).

According to the WHO 2008 report, chronic diseases accounted for almost 60% of deaths in the world and 87% in high-income countries. The proportion of deaths due to chronic diseases is projected to rise up to 69% in 2030 worldwide. In 2014 PAHO stated that annually, almost 4 million people in the Region of the Americas die from chronic noncommunicable diseases, comprising 76% of all deaths. More than one-third of these deaths are premature (occurring before age 70), and most are preventable and can be postponed (WHO, 2013—2019, 2014; PAHO, 2013).

In the United States, chronic diseases and conditions, such as heart disease, stroke, cancer, diabetes type 2, obesity, and arthritis, are among the most common, costly, and preventable of all health problems. For example, in 2012, about half of all adults, 117 million people, had one or more chronic health conditions. One of four adults had two or more chronic health conditions (Ward et al., 2014). Seven of the top 10 causes of death in 2010 were chronic diseases with two of these, heart disease and cancer, together accounted for nearly 48% of all deaths (Centers for Disease Control and Prevention, 2016).

The Triple Aim is a framework developed by the Institute for Healthcare Improvement (IHI) that describes an approach for optimizing health system performance. First introduced in 2007 IHI's Triple aim anchors in the position that new designs must be developed to simultaneously pursue three dimensions (http://www.ihi.org/, last accessed May 15, 2016):

- improving the patient experience of care (including quality and satisfaction);
- improving the health of populations; and
- reducing the per capita cost of health care.

The patient experience component of the Triple Aim encompasses all six dimensions described by Institute of Medicine (safe, effective, patient-centered, timely, efficient, and equitable). It is clear that patient engagement, teams working together local to global, and creating innovative models of care are key strategies for improving the Triple Aim and health. Improving health is a challenge that requires the engagement of partners across the community to address the broader social determinants of health, local to global. Within the United States, a movement in 2014 to add an additional component to the Triple Aim, that of Caregiver Experience, occurred. Leslie Small notes that the precondition for restoring joy and meaning to work is to ensure that the workforce has physical and psychological freedom from harm, neglect, and disrespect. She notes that a health system aspiring to the Triple Aim, fulfilling this precondition must be a nonnegotiable, enduring property of the system (http://www.fiercehealthcare.com/story/why-healthcare-organizations-must-embrace-quadruple-aim/2015—06—04, last accessed May 15, 2016). This aim addresses all caregivers, lay and professional.

Within this scenario, ICT is a fundamental and essential resource to achieve significant results as defined by the Triple Aim and the additional quadruple aim. Expected outcomes of implementing healthcare information systems include quality and service improvement as well as cost reduction and productivity enhancement (Vogel, 2014). Examples of cost savings can range from efficient management of supplies, drugs, laboratory exams to the rational scheduling of surgical rooms, image equipment, and healthcare personal. ICT has the potential to reduce duplication of tests, procedures, and requests, and improve the efficiency to access information and enhance decision-making for diagnosis and treatments.

A second area of benefit comes in the form of improved productivity of clinicians and other staff. With continuing (and at times increasing) constraints on reimbursements, healthcare organizations are continually faced with the challenge of doing more with less. Providing information system support to providers and staff can in many cases enable them to manage a large variety of tasks and data that would otherwise be impossible when using strictly manual/paper-based processes. Interestingly, in some cases, hospital investments in healthcare systems support the improved productivity of staff who are not employed by the hospital, namely the physicians and other healthcare collaborators, and can even extend to payers by lowering their costs. One of the major challenges with introducing a new Health Care Information Systems (HCIS) is that the productivity of users may actually decrease in the initial months of the implementation. With complex clinical applications in particular, learning new ways of working can lead to high levels of user dissatisfaction in addition to lowered productivity (Vogel, 2014).

As mentioned, use of ICT in healthcare can convey clinical, organizational, and financial benefits. Measuring these benefits is variable with some more tangible than others to measure. However, it is important to highlight that financial benefits are not immediately realized after the initial deployment. Many systems that were implemented during the last 40 years are just now able to demonstrate efficiency, maturity, and benefits.

The Meaningful Use initiative of the Office of the National Coordinator for Health Information Technology (ONC) (https://www.healthit.gov/newsroom/about-onc; www.healthit.gov/providers-professionals/how-attain-meaningful-use, last accessed May 8, 2016) introduced a strong strategy to initiate the production of quality indicators from the use of an Electronic Health Record (EHR) (Stage 1) and in the long term (Stages 2 and 3 in years 2014–16) to improve quality, safety, and efficiency that could led to better healthcare outcomes (Degoulet, 2014; HIMSS Analytics, 2013).

Adoption of Meaningful Use has been pervasive in the United States. For example, as of 2015, 95% of all eligible and critical access hospitals demonstrated meaningful use of certified health IT through participation in the Centers for Medicare and Medicaid Services (CMS) EHR Incentive Programs. Ninety-eight percent of all hospitals demonstrated meaningful use and/or adopted, implemented or upgraded (AIU) any EHR as of the end of 2015, and 56% all US office-based physicians (MD/DO) demonstrated meaningful use of certified health IT in the CMS EHR Incentive Programs. Sixteen percent of nurse practitioners (NPs) and less than 2% of physician assistants (PAs) demonstrated meaningful use of certified health IT. Recent statistics note that 48% of all office-based physicians, NPs, and PAs have demonstrated meaningful use of certified health IT. (http://dashboard. healthit.gov/quickstats/quickstats.php, last accessed May 9, 2016).

Information is a key factor to improved health conditions and prevention of diseases. Technology is a fundamental resource that can be used to provide the population with information and knowledge. The range of ICT healthcare strategies addressing information needs comprises the creation of website, portals, social network, and synchronous and asynchronous communication. Many of these strategies are low cost and not complex processes. These tools have brought about a significant change in the daily lives of individuals and families and represent a significant change in healthcare delivery and patient engagement and empowerment. These tools are important supports for health promotion and support the coordinated work of social sectors and the population, consistent with personal, family, and communities' environments. Examples of tools and strategies include Quantified Self Movement (1 trillion sensors), Internet of Things, Social Media (1.8 billion subscribers), and eMobile Health (6 billion cellphones).

Although health promotion, patient participation, development of portals, applications, and social media are examples of how ICT can promote patient engagement, it must be highlighted that patient engagement also means change in culture and care delivery models. It includes a culture that welcomes the participation and collaboration of the patient in the decisions related to his/her healthcare status. Patient collaboration involves a number of elements, including unrestricted communication among collaborators, shared decision-making, mutual respect, and transparent sharing of information and engagement (Sands, 2015). In addition, patient collaboration and engagement also have a closer relationship with patient family and significant others.

Globally, cultural implications must be considered in all patient engagement. Culture definitions vary as well as share similar perspectives and interests within the same culture. An organization must be sensitive to combine common values, vision, policies, attitudes, behaviors and beliefs, as well as different cultures with different defining characteristics.

Patient engagement strategies must be articulated into system-level characteristics that reflect a patient-centered culture where ICT development of different resources and tools should also be based on how to guide an organization toward a more patient-centered culture (Maurer et al., 2012; Guide to Patient and Family Engagement in Hospital Quality and Safety, 2016). As pointed out by Collier (2015), the literature contains descriptions of why patient engagement matters and what defines a patient-centered environment. The evidence of the impact of patient engagement on costs, outcomes, and experience is growing daily and several experiences have shown that all the information, evidence, and persuasion available will not translate to change in behavior unless individuals can articulate and demonstrate what patient engagement means to them on a personal level. It is necessary to demonstrate through words, behaviors, and actions that patient partnership matters to achieve success in healthcare status.

As an exemplar, Center for Advancing Health — CFAH (2010) defines patient engagement as "actions individuals must take to obtain the greatest benefit from the health care services available to them." As can be observed, the definition emphasizes the behavior of the individuals as a stronger component for a successful healthcare outcome rather than influence of the clinical providers. However, as mentioned before, outcomes in healthcare are complex and depend on many different factors, including patient, family, providers, environment, resources, national, regional and international (such as WHO) institutes, as well as individual characteristics such as age, literacy, habits among others.

Moreover, it is valuable to emphasize that engagement is not synonymous with compliance. Compliance means that an individual obeys a directive from a healthcare provider. Engagement signifies that a person is involved in a process through which the patient harmonizes robust information and professional advice with their own needs, preferences, and abilities in order to prevent, manage, and cure disease (Center for Advancing Health — CFAH, 2010).

Consider a small scenario where patient and family (understood as anyone the patient deems as family regardless of biological association) are engaged. There are many levels describing this engagement; this example describes three levels that are proposed by Carman et al. (2013):

*Level 1*: Patient and providers make decisions based on the medical evidence, patients' preferences, and clinical judgment.
*Level 2*: Organizational design and governance—healthcare organizations reach out to patients for input to ensure that they are being as responsive as possible to patients' needs.
*Level 3*: Policies—definitions exist to assure that patients are involved in the decisions that communities and society make about policies, laws, and regulations in public health and healthcare.

The National Academy of Medicine defines Patient Family Engagement as, "a set of behaviors by patients, family members, and health professionals and a set of organizational policies and procedures that foster both the inclusion of patients and family members as active members of the health care team and collaborative partnerships with providers and provider organizations." (Olsen et al., 2011). The collaborations among all components involved in healthcare delivery or maintenance are fundamental to achieve better results, improve health of the population, achieve better patient experience of care, and reduce costs as well as assure safety.

The importance of patient engagement, experience, and satisfaction is growing as patients seek to play a larger role in their own care. If the patient can understand his/her health condition, know the symptoms, knows why they are taking medication, and understands how to implement necessary lifestyle changes, the chances of getting and staying healthy are significantly improved (Blanton, 2015).

Within a breadth of potential of patient and family engagement related to the dimensions, definitions, and scope, a consensus has emerged in two aspects: (1) patient engagement improves healthcare outcomes; and (2) technology can support engagement (HIMSS North America, 2014).

People actively engaged in their health are more likely to stay healthy and manage their conditions, follow treatment plans, eat right, exercise, and receive health screenings and immunizations (Hibbard and Cuningham, 2008). Conversely, patients without the skills to manage their healthcare experience cost up to 21% more than patients who are highly engaged in care (James, 2013). It can be inferred that when patients seek adequate information, change habits, and follow clinicians advises, the chances of obtaining better results and better quality of living are higher.

However, evaluating and quantifying patient engagement is challenging? Which indicators and strategies can support providers to consider that patient behavior can be sufficient to improve health status? A score to measure the degree to which someone sees himself or herself as a manager of his or her health and care was proposed by Hibbard of the University of Oregon, United States. In their study the relationship between patients' activation scores and their healthcare costs at a large healthcare delivery system in Minnesota, United States was analyzed (Hibbard and Greene, 2013). The results showed that those with the lowest activation scores, i.e., people with the least skills and confidence to actively engage in their own healthcare, incurred costs that averaged 8–21% higher than patients with the highest activation levels, even after adjusting for health status and other factors. The authors concluded that patient activation scores were shown to be significant predictors of healthcare costs.

Patient engagement requires literacy. Individuals must know what should be known, where the best information is available, what is reliable, and where to find information as needed. Providers should welcome the opportunity to partner with their patients and encourage them to access health information. As suggested by Sands (2015) this can be done regularly by asking patients about their use of outside information, by looking up information with patients, and by helping them find useful and adequate information, since the portals, websites, and all resources available are not always sufficient to answer specific questions and clarify doubts. Health literacy and health education are highly complex themes that demand clear articulation among all subjects involved. Further reading is suggested on exploring the complexity of health literacy are provided in this chapter.

ICT represents a strong power to support healthcare, education, management as much as it is presents at all human activities. Although healthcare delivery has traditionally been focused on acute conditions, it is necessary today to design models and systems according to the contemporary conditions of living, and being health, recovering health, dealing with chronic diseases, facing endemics, and having a dignified end of life.

Seeing patient care as a collaboration between providers and patients around patient's health and life conditions enables us to understand how to transform healthcare by focusing on nonvisit-based care, patient engagement, and new models of care delivery. These elements can leverage ICT and healthcare information technology to improve the population's heath, reduce costs, and improve the experience of care for all stakeholders, thus achieving the Triple Aim, as proposed by Institute for Healthcare Improvement (2016) and Berwick et al. (2008).

## DIGITAL HEALTH COMMUNITIES

The advent of the Internet with democratization of health information and knowledge has contributed to the exponential growing of social media. This is a key example of how ICT can be used to improve healthcare. However, patient engagement is not just the use of ICT or just the development of computer and mobile applications and implementation of online patient portals, websites, social media groups, and e-messaging.

It is a fact that the Internet technology and social media are making it easier than ever for consumers to find timely, personalized healthcare information online. Previously connected mainly through email discussion groups and chat rooms, patients are now able to build more sophisticated virtual communities that enable them to share information about treatments and support, and build online personal networks of friends (Ba and Wang, 2013).

Many examples of resources where patients, mainly for individuals with chronic diseases, are available worldwide. Patientslikeme.com is a platform where patients can share information to improve their lives through providing resources for chatting on the website, blogging about their illness, and supporting each other with recommendations.

To create digital communities, digital inclusion must have be established as a requirement. Citizens must have the ability to access and to use information and communication technologies, not just mobile phones, but also computers, Internet, and available software. Thus digital inclusion encompasses searching for relevant content and services in the Internet and demands consequent training and computer and health literacy for the effective use of information and communication technologies (Institute of Museum and Library Services, 2012). Digital inclusion is the underpinning factor for building healthy digital communities across sectors such as economics and workforce development, education, healthcare, public safety and emergency services, civic engagement, and social connections (Institute of Museum and Library Services, 2012). Although each community will have different priorities, the fundamental needs are the same and include:

- high-speed infrastructure to carry Internet traffic and connect households to businesses and community institutions;
- help people learn to navigate technology and provide a safe environment for doing so;
- maintain public options for those who are unable, cannot afford, need temporary access, or for whom it does not make sense to maintain private access;
- remove barriers to access for people with disabilities or facing other obstacles to access; and
- address the special needs and requirements of key sectors that drive the local economy.

To achieve success, a digital community must engage social sector, government, vendors, stakeholders, and special interest groups, like chronic disease patients' groups such as Parkinson and diabetes.

Medline Plus offers a section called "Videos and Cool Tools" that contains interactive tutorials of diseases and conditions, tests and diagnostic procedures, surgery and treatment procedures, and prevention and wellness. The section also offers animated anatomy videos, surgery videos, and games to help patients learn and assess their knowledge. The following table summarizes additional resources.

| Taking Charge of Your Health | University of Minnesota's Center for Spirituality and Healing and the Life Science Foundation have created a website that provides reliable information about alternative and integrative health practices. Available at: www.takingcharge.csh.umn.edu |
|---|---|
| Cancer.gov | Hosted by the National Cancer Institute, the site provides information on different types of cancers, treatments, clinical trials, and more. One section includes integrative therapies. The site is for patients and their families, as well as healthcare providers. Available at: http://cancer.gov/ |
| MedlinePlus | Commercial-free, reliable, free health information from the National Library of Medicine including health topics, drug information, health news, medical encyclopedia and dictionary, interactive tutorials, and more. Available at: http://www.nlm.nih.gov/medlineplus/ |
| Kidshealth | The Nemours Foundation's Center for Children's Health provides sites for parents, children, and teens containing current information (in English and Spanish) about child development, nutrition and fitness, preventive health care, and diseases and conditions. Available at: http://kidshealth.org/ |
| NIH Senior Health | From the National Institutes of Health, this site provides information on many topics from cancer to mental health and is specifically designed for older adults, with features such as enlarged text and audio. Available at: http://nihseniorhealth.gov/ |
| Ethnomed | A joint program of the University of Washington Health Sciences Libraries and Harborview Medical Center, this site includes patient education materials on topics ranging from immunizations to asthma to smoking and more. It provides information in several languages, including Hmong and Somali. Available at: http://ethnomed.org/ |
| Healthy Roads Media | This site was created by Healthy Roads Media with support from the National Library of Medicine and others and provides education materials in a number of languages and a variety of formats (including videos, handouts, and audio). Available at: http://healthyroadsmedia.org/ |

## CHALLENGES AND OPPORTUNITIES

Among all healthcare strategies being implemented by countries, ICT is a priority. As a priority to reach high volume and quality of information for evaluation and healthcare planning, most countries must invest on affordable technology and deploy resources that enable the provision of provide care and expertise at a distance. ICT can provide clinical decision support tools and systems with online access to research evidence, guidelines and patient information, targeted educational interventions using EHRs and web-based portals. In addition, ICT is mandatory for acquisition, storage, and management of clinical and administrative data of patient and population levels. (OECD, Improving Health Sector Efficiency, 2010; European Health Telematics Association EHTEL, 2008).

However, effective operation requires strong leadership at the national, regional, and organizational level. The government regulatory and supervisory role is imperative for successful implementation. The deployment of ICT in health services is a major infrastructural investment and the adaptation by health service providers to this infrastructure should be supported (Kem, 2011).

Despite the challenges and opportunities, it is important to note the worth of health and education as strong pillars to develop. Health can be improved by access of information and educational programs. Education can be developed and enhanced by healthy population input; this creates a renewing circle. Geographically, distance, rural communities, developing countries, and health disparities are factors that stimulate the development of e-health resources to reach vulnerable population.

The ICT in healthcare can provide opportunities to better leverage of resources and improve effectiveness of national and international health systems. As the costs in healthcare reach high levels, new delivery models and technology could increase life conditions to promote healthy living. Technology can bring information to the point of need and stimulate consumers to take responsibility of promoting a healthy environment and personal behaviors (Frieden, 2015). For several countries and populations, ICT adoption has been slow due to cost, infrastructure, conception models, architecture, integration, usability, and public policies implementation. Adoption and deployment are also dependent on training and education. It is mandatory to influence government leaders about the importance of education and preparation as fundamental to ensuring that patients and families can engage effectively in the ways that they want and need to achieve optimum health status as possible. This includes education and preparation related to their own health and healthcare and preparation to partner with clinicians and healthcare leaders to shape how care is organized and delivered. A key purpose of education and preparation is giving patients and families the skills, confidence, and authority to partner—to the degree that they want—in interactions and healthcare decision-making at all levels and to provide self-care and manage illness and chronic disease effectively. Because patients and families are diverse in their desire and ability to engage, it is important to consider how we can tailor efforts to meet patients and families where they are, address specific needs and concerns, and best facilitate their engagement.

An important factor to success is relevance. Globally, some diseases are considered universal and they affect populations independent of race, gender, economic development, and others. These are relevant for the healthcare system, for the economy and governance, and for the individuals. Having digital communities that can cross boundaries and can reach different nations will promote a resources for sharing experiences and for achieving better outcomes and will play a significant role on the management diseases for those individuals. However, beyond having access to technologies, individuals, businesses, and institutions need to understand the relevance and benefits of using technology to achieve educational, economic, and social goals. Awareness of its potential benefits creates the motivation to master technology skills and learn more about effectively using broadband. These goals aim to help communities raise awareness of the value of technology and generate interest in adopting its use (Institute of Museum and Library Services, 2012)

As mentioned before, access to the Internet is also not enough to be affordable for individuals, communities, and provides significant service to expand digital communities and resources on healthcare management. Government bodies, vendors, academia, private institutes must work together to decrease costs of digital access.

## FINAL CONSIDERATIONS

Internet has changed information on health acquisition. Internet has changed daily activities, habits and empowered citizens as much as diminished barriers and frontiers across countries, regions, and communities. Technology has been developed and deployed to serve the human needs on information, interaction, and support online communities to congregate citizens and peers for sharing experiences, information, and knowledge.

As healthcare providers, we must also be engaged and make sure to be involved on those activities. We must be sure all ICT resources represents strategies and solutions to empower citizens, patients, and members of health digital communities. We must work together with the patients and for the patients to achieve better solutions and better outcomes, improving health quality and life conditions for all.

## References

Ba, S., Wang, L., 2013. Digital health communities: the effect of their motivation mechanisms. Decis. Support Syst. 55 (4), 941–947, <http://dx.doi.org/10.1016/j.dss.2013.01.003>.

Berwick, D.M., Nolan, T.W., Whittington, J., 2008. The triple aim: care, health, and cost. Health Affairs 27 (3), 759–769, <http://dx.doi.org/10.1377/hlthaff.27.3.759>.

Blanton, K., 2015. The patient engagement prescription. N. C. Med. J. 76 (3), 157–160.

Carman, K.L., Dardess, P., Maurer, M., et al., 2013. Patient and family engagement: a framework for understanding the elements and developing interventions and policies. Health Aff. (Millwood) 32 (2), 223–231.

Center for Advancing Health – CFAH, 2010. A New Definition of Patient Engagement: What Is Engagement and Why Is It Important? CFAH, Washington, DC, USA. Available on: www.cfah.org. Accessed April 8, 2016.

Centers for Disease Control and Prevention. Death and Mortality. NCHS FastStats Web site. <http://www.cdc.gov/nchs/fastats/deaths.htm> (accessed 10.05.16).

Collier, M.S., 2015. Culture change to promote patient engagement: it isn't how....it's who. N. C. Med. J. 76 (3), 168–170.

Degoulet, P., 2014. Hospital information systems. In: Venot, A., Burgun, A., Quantin, C. (Eds.), Medical Informatics, e-Health. Springer Paris, Springer-Verlag France, pp. 289–313.

European Health Telematics Association (EHTEL), 2008. Sustainable telemedicine: paradigms for future-proof healthcare, a briefing paper. <http://www.ehtel.org/forum/tasks-sources/task-force-sustainable-telemedicine-and-chronic-disease-management/ehtel-briefing-paper-sustainable-telemedicine-paradigms-for-future-proof-healthcare-1/files/ehtel-briefing-paper-sustainable-telemedicine.pdf>.

Fernandez, A., Oviedo, E. (Eds.), 2011. e-Health in Latin America and the Caribbean: Progress and Challenges. ECLAC, Santiago, Chile.

Frieden, T.R., 2015. The future of public health. N. Engl. J. Med 373, 1748–1754.

Guide to Patient and Family Engagement in Hospital Quality and Safety. Agency for healthcare research and quality website. <http://www.ahrq.gov/professionals/systems/hospital/engagingfamilies/guide> (accessed 08.04.16).

Hibbard, J.H., Cuningham, P.J., 2008. Research Brief No. 8: How Engaged Are Consumers in Their Health and Health Care, and Why Does It Matter? Center for Studying Health System Change, Washington, DC.

Hibbard, J.H., Greene, J., 2013. What the evidence shows about patient activation: better health outcomes and care experiences; fewer data on costs. Health Aff. 32 (2), 207–214.

HIMSS Analytics, 2013. Available at: <http://www.himssanalytics.org/home/index.aspx> (accessed 01.06.16).

HIMSS North America. The State of Patient Engagement and Health IT, August 12, 2014.

Institute for Healthcare Improvement, 2016. The IHI Triple Aim. Available at: <http://www.ihi.org/engage/initiatives/tripleaim/Pages/default.aspx> (accessed 10.04.16).

Institute of Museum and Library Services, University of Washington Technology & Social Change Group, International City/County Management Association, 2012. Building Digital Communities. Institute of Museum and Library Services, Washington, DC. Available at <http://tascha.uw.edu/digital-inclusion-framework>.

James J., Health policy brief: patient engagement. Health Aff., February 14, 2013.

Kem, Z.G., 2011. Searching for an integrated strategy in chronic disease management: a potential role for information and communication technologies. Manage. Health 15 (4).

Maurer, M., Dardess, P., Carman, K.L., Frazier, K., Smeeding, L., 2012. Guide to Patient and Family Engagement: Environmental Scan Report. AHRQ Publication No.12–0042-EF. Agency for Healthcare Research and Quality, Rockville, MD.

Neto, G.V., 2014. A evolução da assistência à Saúde no mundo e no Brasil até o SUS. In: Neto, G.V., Malik, A.M. (Eds.), Gestão em Saúde. Guanabara Koogan, Rio de Janeiro, pp. 3–14.

OECD, Improving Health Sector Efficiency, 2010. The role of Information and Communication Technologies, OECD Health Policy Studies. <http://ec.europa.eu/health/eu_world/docs/oecd_ict_en.pdf>.

(Chapter 4) Engaging patients to improve science and value in a learning health system. In: Olsen, L.A., Saunders, R.S., McGinnis, M., Olsen, L.A., Saunders, R.S., McGinnis, M.

(Eds.), Patients Charting the Course: Citizen Engagement in the Learning Health System—Workshop Summary. National Academies Press, Washington, DC.

PAHO, 2013. Innovative care for chronic conditions. Organizing and Delivering High Quality Care for Chronic Non Communicable Diseases in the Americas. PAHO, Washington, DC.

Sands, D.Z., 2015. Health care as collaboration: ICT enabling the role of the patient in Health Care Transformation. CGI.BR/CETIC.BR. ICT in Health 2014 – Survey on the Use of Information and Communication Technologies in Brazilian Healthcare facilities. Coordinator: Alexandre Barbosa, São Paulo, pp. 179–186.

Vogel, L.H., 2014. Management of information in health care organizations, <http://dx.doi.org/10.1007/978-1-4471-4474-8_14>. In: Shortliffe, E.H., Cimino, J.J. (Eds.), Biomedical Informatics, 1. Springer-Verlag, London.

Ward, B.W., Schiller, J.S., Goodman, R.A., 2014. Multiple chronic conditions among US adults: a 2012 update. Prev. Chronic Dis. 11, 130389, <http://dx.doi.org/10.5888/pcd11.130389>.

WHO, 2013–2019, 2014. Action Plan for the Global Strategy for the Prevention and Control of Non communicable Diseases: Prevent and Control Cardiovascular Diseases, Cancers, Chronic Respiratory Diseases and Diabetes. WHO, Geneva.

## Further Reading

Westra, B., Weinferter, L., Delaney, C., 2014. Integrative nursing and health literacy. In: Kreitzer, M.J., Koithan, M. (Eds.), Integrative Nursing. Oxford University Press, Cary, NC, pp. 445–455. (Chapter 32).

Oldenburg, J., 2013. Engage! Transforming Healthcare Through Digital Patient Engagement. Healthcare Information and Management Systems Society (HIMSS), Chicago, USA, ISBN 13978–1–938904–39–4.

# CHAPTER 11

# Ethical and Transborder Issues

**E. Massad**

University of São Paulo, São Paulo, Brazil

## CONTENTS

## INTRODUCTION

Is the World prepared for the next "Big One" pandemic? Unfortunately, the honest answer should be a resounding NO! Not yet, at least. Of course this will very much depend on what, where and when next big one will come. You can make your bet. My hunch is that it will be a zoonotic respiratory pathogen with a long and infective incubation period (you guessed it; HIV was almost there but its transmission is too "complicated"). Experts propose that the next pandemic will be caused by a pathogen with high "intrinsic evolvability" (Quammen, 2013), which is a pathogen that mutates very quickly and/or recombines its genetic material when replicates. But the truth is that from time to time the world is put on high alert, triggered by some exotic and frequently unknown infectious disease spread in less than 24 hours by travelers of international routes to previously uninfected regions (Stannard, 1993). The most notorious example is perhaps the Black Death of the 14th century, which decimated from a quarter to a half of European population (Massad et al., 2004; Bossak and Welford, 2009). Many centuries later, the Spanish Flu, which killed between 50 and 100 million individuals worldwide, started in an American army barrack in the United States in 1918, and rapidly spread by travelers to others areas of the world (Caley et al., 2007; Massad et al., 2007). Almost 100 years later, SARS frightened many countries due to its potential spread by infected travelers (Wilder-Smith and Freedman, 2003; Massad et al., 2005a). The swine flu pandemic (H1N1) of 2009 is another example of the dangers of international spread of communicable diseases (Khan et al., 2010; Massad et al., 2010). The recent outbreak of Ebola is the most recent example of the risk of a new and, in this case, frequently fatal disease, posed by individuals traveling from

**Global Health Informatics.**

infected to uninfected areas of the world (Gomes et al., 2014; Pandey et al., 2014). And now came Zika virus...

In this chapter I address the ethical issues of transborder problems, emphasizing the moral responsibility of decision-makers and the desperate need of new information technologies to help curbing the imminent risk of mass killing of emergent and/or reemergent infectious diseases. First, I present the bare essentials of bioethics, its definitions and key components. Next I address the concept of Macroethics, the moral responsibility of nations, international organisms or group of individuals with the power of deciding who will be entitled to receive appropriate treatment or prevention. This contrasts with the classical, but not-yet formalized, concept of Microethics, the moral responsibility of individual agents. Following this section I propose, through some real life examples, how information technology in the form of mathematical and computer models can be decisive on the just distribution of preventative or curative measures that can mitigate the impact of emergent and reemergent infectious diseases and, perhaps more important, the unavoidable nightmare of the "next big one."

## BIOETHICS—THE BARE ESSENTIALS

Bioethics is the branch of ethics that deals with ethical evaluation that include the morality of all actions that might help or harm organisms capable of feeling fear. It was firstly proposed as an applied field of ethics in the 1920s by Fritz Jahr (Rinčić and Muzur, 2012) and was popularized by Van Rensselaer Potter (Beuchamp and Childless, 2001) who proposed a "global ethics," linking biology, ecology, medicine, and human values. Potter proposed bioethics as a way to attain the survival of both human beings and other animal species. Bioethicists are typically concerned with the ethical questions that arise in the relationships among life sciences, biotechnology, medicine, politics, law, and philosophy.

The fundamental principles regarding the use of human beings in medical research were announced in the Belmont Report (DHEW, 1978) and are: autonomy, beneficence, and justice. Although aimed at medical research, these principles have influenced the thinking of bioethicists across a wide range of issues. Other values later included are nonmaleficence, human dignity, and the sanctity of life.

As a branch of the philosophical discipline of ethics, bioethics should concern with what people value (conventions) and, therefore, how people should behave and what they ought to believe (Briggle and Mitcham, 2012). As mentioned by Briggle and Mitcham (2012), however, "people create, not discover, values. In this way the world of science (facts) is divorced from ethics (values)" (Briggle and Mitcham, 2012, p. 34). This conflict between thinking (science) and conformity (values) is, perhaps the reason why scientist tend to view the issues addressed by ethics (and bioethics in particular) with contempt. Technology, in contrast, as applied science to solve real-world problems, cannot avoid the moral responsibility of its applications. Hence the necessity of a normative ethics, consisting of theories about what makes persons' states of affairs and actions, virtuous or vicious (virtue ethics), good or bad (consequentialist ethics), and right or wrong (deontological ethics). Bioethics added a fourth theory, the ethics of care, which focus on interpersonal relationships. According to the ethics of care, actions are considered right and outcomes as good insofar as they justify these relationships.

## MACROETHICS

Macroethics is a recently introduced concept (Vallero, 2007) and addresses the moral responsibility of whole communities, like governments, international bodies, and other nongovernmental organizations in which a few decide the destiny of many. It differs from Microethics, the moral responsibilities of individuals. Macroethics has been discussed moral dilemmas in biotechnology (Vallero, 2007), genomics (Bassaingthwaighte, 2003), and engineering (Herkert, 2004). In this chapter the concept of Macroethics is applied in the field of large-scale pandemic threats of emerging and reemerging infectious diseases. Information technology in the form of mathematical and computational models is presented as an invaluable tool in helping deciding, at a macro level, who, when, and how finite resources like vaccines or medicines should be justly distributed such that the utilitarian principle of the greatest good to the greatest number of people is achievable.

## BIOETHICS OF INFORMATION TECHNOLOGY: KEY COMPONENTS

Assuming, as in Mason (1995), that an ethical issue arise whenever one party in pursuit of its goal behaves in ways that affects the ability of another party to pursuit its goals, in the bioethics of information technology context it is interesting to try to answer the following question: *When a moral agent decides to change the state of information or information technology how this will impact a human system?* To answer this question, it is necessary to identify the key components involved in the ethical issues related to IT in general. They are:

1. The Agents: individuals (Microethics) or group of individuals (Macroethics) whose behavior changes the informational status and/or make decisions:
2. The Acts: the very behavior relating to informational status and/or decision-making;

3. The Results: consequences of the changing in informational status and/or decision-making; and

4. The Stakeholders: individuals/populations who suffer the consequences (results) of the agents' actions.

In addition, two ethical perspectives should be considered in the bioethical analysis of IT, namely, the Deontological (duties) and the Axiological (values) perspectives.

In Deontological ethics an action is considered as right when it conforms to moral principles manifested in duties rather than in terms of virtue (good consequences). Axiological ethics, in turn, addresses the values such as the pursuit of happiness, the pursuit of virtues or the pursuit of justice. Both deontological and axiological perspectives address such questions as *who should decide?*; *who should benefit?*; *how should the decisions be made?*; and *how can the issues be prevented from arising in the future?*

In the following sections I exemplify the concepts above with the case of a current transborder issue that is causing a major concern to health authorities around the globe: emerging and reemerging infectious diseases.

## BIOETHICS OF INFECTIOUS DISEASES

Much of the work of early bioethics consisted in the discussion of dilemma-cases (Battin et al., 2009). The core initial areas of concern of bioethics were experimentation with human subjects, genetics, organ transplantation and artificial organs, the ethics of death and dying, and reproductive ethics (Jonsen, 1998). In contrast, for almost all of bioethics' history, infectious diseases have simply been left out. The reasons for the virtual neglect of infectious diseases by the bioethics community have been extensively discussed in the recent literature. Here we quote the work by Selgelid (2005), who suggested why ethical issues associated with infectious diseases have not been more prominent in bioethics literature basing on six arguments: (1) bioethics was born at a time of important advances of medical technology and thus has largely focused on these; (2) the birth and initial development of medical ethics occurred at a time when it was believed that infectious diseases was almost completely conquered by medicine (this lead the Surgeon General of the United States to the limit of hubris in his (in) famous pronouncement of 1967: "The book of infectious diseases must be closed. . ."); (3) the biased point of view that infectious diseases are problems of "others"; (4) the difficulty in carrying out research on infectious diseases in the developing world due to fact that they are so empirical and interdisciplinary; (5) the ethical questions posed by infectious diseases do not appear to pose the kind of "deep" philosophical questions that academic ethicists of the developed world are interested in; and (6) debate in bioethics has, to a large extend, been hijacked by religion.

According to Battin et al. (2009), canonical accounts of principles in bioethics (Beuchamp and Childress, 2001), autonomy, nonmaleficence, beneficence, and justice have not been tested in infectious diseases examples.

Bioethics has focused on protecting individuals but infectious diseases challenges bioethics to develop new ways of understanding autonomy and related concepts (Battin et al., 2009) as it is difficult to strike a balance between the utilitarian aim of promoting public health, on the one hand, and the libertarian aims of protecting privacy and freedom of movement, on the other (Selgelid, 2005).

Infectious diseases deeply involve issues of justice (Selgelid, 2005) and at the same time, raise difficult philosophical and ethical questions of their own. The measures required for their control and to protect individuals and society involve surveillance and notification of authorities (against privacy), mandatory testing, vaccination or treatment (against autonomy), isolation of individuals, quarantine of regions, or restriction to travel (against liberty of movements).

In this chapter it is argued that the use of mathematical and computer models for designing interventions that immunization and/or treatment strategies should be an ethical priority since they help minimize the conflict between the interests of individuals and those of society at large. The use of such tools to estimate the risk of spreading these infections from source countries/regions is also exemplified by the cases of Ebola and Zika virus. Final comments deal with the current case of urban yellow fever (YF) in Angola.

## MACROETHICAL DILEMMAS IN PUBLIC HEALTH

With the increasing emergence and resurgence of infectious diseases around the world and the consequent implementation of public health actions to control them, conflicts between the perceptions of rights (individual vs collective) are bound to grow (Velimirovic, 1987). This kind of conflict is as old as the first effective public health actions, such as inoculation against smallpox. In 1760 the great Swiss mathematician Daniel Bernoulli set out to calculate the increase in average life expectancy that would result from inoculation against smallpox and came up with a figure of 2 years (Dietz and Heesterbeek, 2002). Those who resisted inoculation questioned whether most people were actually prepared to run the significant risk of immediate death, one in a hundred, to gain only 2 years of life (Valleron, 2000).

Another emblematic instance is that involving the US Supreme Court case of *Jacobson v. Massachusetts*. One hundred years ago, this case established a four-part test for evaluating claims of individuals that government public health authorities were infringing on their liberty by imposing compulsory vaccination (Fabro and Jacobson, 1970; Parmet et al., 2005; Blum and Talib, 2006).

No remedial intervention of any sort is absolutely risk-free. Even the commonest drug may carry some degree of risk to a fairly large number of users, and a possibly life threatening risk to an exceedingly small number of persons who are hypersensitive to it. Any public health or medical judgment must be made on the basis of balancing the values and the problems of one procedure against those of another procedure and against the risk of doing nothing at all (Melnick, 1978). Vaccination, a deliberate attempt to protect humans and animals against diseases, is not different.

Since the time of Edward Jenner, vaccination has controlled 10 major human diseases, at least in some parts of the world (Plotkin and Plotkin, 1994): smallpox, rabies, diphtheria, tetanus, YF, pertussis, poliomyelitis, measles, mumps, and rubella.

Unfortunately the development of resistance against antibiotics and the environmental disruption and political turmoil of last century determined a growing difficulty to control old pathogens and the emergence of new ones. In addition, serious side effects of old and new vaccines started to be reported in the last few years. The reversal of vaccine to wild type of attenuated virus vaccines like the Sabin vaccine against poliomyelitis and the deaths associated to YF and smallpox vaccines are causing great concern among public health (and more recently, military) authorities.

Voluntary vaccination policies for childhood diseases present parents with a subtle challenge: if a sufficient proportion of the population is already immune, either naturally or by vaccination, then even the slightest risk associated with vaccination will outweigh the risk from infection. As a result, individual self-interest might preclude complete eradication of a vaccine-preventable disease. Formal game theoretical analysis of this problem may lead to new insights that help to explain human decision-making with respect to vaccination. Increases in perceived vaccine risk will tend to induce larger declines in vaccine uptake for pathogens that cause more secondary infections (such as measles and pertussis). After a vaccine scare, even if perceived vaccine risk is greatly reduced, it will be relatively difficult to restore pre-scare vaccine coverage levels (Bauch and Earn, 2004).

## THE "CLASSICAL" APPROACH TO VACCINE OPTIMIZATION

The success of vaccination programs relies on a concept known as herd immunity, according to which protection of a whole community is achieved through attaining a vaccination level above a certain threshold. It is for this reason that vaccination does not require 100% compliance to be effective. Therefore, so long as this threshold is attained, those who refuse to be vaccinated are indirectly protected through the immunization level attained by a minimum level of the community. These "suckers" are violating the microethical principle of justice!

The first proposal to optimize vaccination schedules is probably due to Anderson and May (1991) and it is based on the concept of the "basic reproduction number." A central parameter related to the intensity of transmission of infections, the basic reproduction number ($R_0$), was first defined by Macdonald (1952) as the number of secondary infections produced by a single infective in an entirely susceptible population (Massad et al., 1994). It is a function of the transmission parameter $\beta$, a composite parameter combining the number of potentially infective contacts and the probability that those contacts will generate a new infection, and the removal rate $\mu$, the sum of the rates of removal from infection by recovery, natural death, and the additional death due to the infection. The basic reproduction number, $R_0$, is the ratio between $\beta$ and $\mu$, i.e.:

$$R_0 = \frac{\beta}{\mu} \tag{11.1}$$

Related to the basic reproduction number is the effective reproduction number $R(t)$, the product of $R_0$ and the proportion of susceptible at each instant of time, $x(t)$, i.e.:

$$R(t) = R_0 x(t) \tag{11.2}$$

Now, if a proportion $p$ is successfully immunized by a 100% efficient vaccine, the proportion of susceptible remaining after vaccination will be $1 - p$, and therefore:

$$R(t) = R_0(1 - p) \tag{11.3}$$

To eradicate the infection, Eq. (11.3) must be less than unit, hence, the critical proportion to be vaccinated, for eradication to be attained—the herd immunity level, is

$$p_c = 1 - \frac{1}{R_0} \tag{11.4}$$

For H1N1 influenza virus, for instance, estimates of $R_0$ vary from 1.3 to 1.7 (Yang et al., 2009). Therefore the critical proportion to be vaccinated ranges from 23% to 41%. However, one should be aware that these estimations are valid for the so-called homogenously mixing assumption (Anderson and May, 1991), according to which all susceptibles have the same probability of contacting all infectives, an unlikely situation for the case of the current flu pandemic. Hence a more sophisticated approach for the estimation of the optimal proportion of susceptible to be vaccinated is needed.

# NONINNOCENT VACCINES

As mentioned above, there is no public health intervention totally devoid of inconveniences. In the case of vaccines, side effects varies from the nuisance of going to the vaccination post, to the pain of injection, fever, mild local hypersensitivity reactions, serious general reactions, to the end of the spectrum with death.

From the herd immunity point of view, the best strategy for the individual would be not to be vaccinated while all other members of the community are vaccinated. The individual would avoid all the inconveniences/risks related to the vaccine and would be indirectly protected at the same time. This strategy is at a frontal conflict with the interests of the community, which would optimize its protection by vaccinating the maximum number of individuals, if possible all.

In this section I present the case of three specific, noninnocent vaccines, namely, smallpox, poliomyelitis and YF, and propose an optimal strategy based on game theory, which is aimed to minimize the number of individuals presenting side effects and infection at the same time.

Smallpox has reemerged recently as a public health issue. Because of the lack of information on smallpox transmission in contemporary populations, mathematical and computer modeling has an especially important role to play in policy development. Suggestions for vaccination policy have ranged from preemptive mass vaccination (Kaplan, Craft, and Wein, 2002) to post-outbreak ring vaccination (Halloran et al., 2002). The most serious adverse effects of smallpox vaccination are encephalopathy and encephalomyelitis, with death probability estimated in around 16 per million doses (Henderson and Moss, 1999).

YF was one of the most feared lethal diseases before the development of an effective vaccine (Monath, 2001). Even today, about 200,000 new cases are reported every year in tropical regions of Africa and South America (see "Coda" section). The main reservoirs for the YF virus are some species of monkeys, and transmission occurs through the bites of infected mosquitos. In the sylvatic cycle the main vectors are mosquitoes from the genera *Haemagogus* and *Sabethes*, whereas the urban cycle involves men and mosquitos of the genus *Aedes*, in particular *Aedes aegypti* and *Aedes albopictus*. The bridge between the sylvatic and the urban cycles depends on humans that go to the sylvatic areas for leisure or work, eventually returning to the urban areas carrying the YF virus.

The disease can be prevented by a live attenuated vaccine prepared from the 17D strain of YF virus, which induces seroconversion in more than 95% of recipients and provides immunity for 30 years or longer (Poland et al., 1981). YF vaccine has been incorporated into routine vaccination programs in South America, but in Africa, coverage rates are low (Martin et al., 2001). Vaccination policies have ranged from preemptive mass vaccination to post-outbreak ring vaccination.

Unfortunately, in recent years, a growing number of death associated with the YF vaccine has been reported in the literature (Martin et al., 2001; Vasconcelos, 2001, Chan et al., 2001; CDC, 2002). The risk of fatal adverse events associated with the 17DD YF vaccine used in Brazil was estimated by Struchiner et al. (2004). Their estimation varied from 0.017 to 12.071 fatalities per million doses administered.

Inactivated (Salk) poliovirus vaccine referred to as IPV has been in use since 1955. Live attenuated (Sabin) trivalent poliovirus vaccine or OPV has been the vaccine of choice in most countries since 1963. Mass immunization has been most successful, and wild-type poliovirus transmission has been eradicated from North America and most industrialized countries in the Western hemisphere in 1991 (WHO, 2003; CDC, 2003). Global eradication of poliomyelitis is within reach and is likely to be attained within the next decade (WHO, 2003, 2015).

The risk of OPV-associated poliomyelitis has been estimated by the World Health Organization at between 0.5 and 3.4 cases per million doses in susceptible children (WHO, 2003, 2015), and this has reestablished a role for the use of the inactivated vaccine and a new strategy for immunization with IPV in the United States and elsewhere (CDC, 2003).

## NASH EQUILIBRIUM

To solve the conflict between individual self-interests and the population interests related to vaccination, Bauch et al. (2003) devised a vaccination game that takes account of the risk associated to the vaccine and that related to the disease, i.e., people not vaccinated by self-interest can catch the infection in the case of an outbreak and suffer the consequences. The game seeks an optimum strategy that minimizes the total number of deaths both by the vaccine and the disease. We present those authors' results for smallpox, the results for YF from Massad et al. (2005b) and present a new analysis for poliomyelitis vaccine.

The most commonly solution concept in the game theory is that of Nash equilibrium (Osborne and Rubistein, 1994). This notion captures a steady state of the play of a strategic game in which each player holds the correct expectation about the other players' behavior and acts rationally. It does not attempt to examine the process by which a steady state is reached.

The concept of Nash equilibrium is so named after for the mathematician, economist, and Nobel Prize winner John Nash, whose work in the late 1940s and early 1950s established this solution method for noncooperative games (Dixit and Skeat, 1999). At the Nash equilibrium, each player must be satisfied with the strategy choice she has made, given what other players have chosen.

Bauch et al. (2003) vaccination game is a population game, meaning that the payoff to an individual choosing a particular strategy depends on the average behavior of the population. The two basic strategies are "vaccinator" (obtain preemptive vaccination) and "delayer" (decline preemptive vaccination but seek vaccination in the event of an attack). For any strategy, the payoff to an individual is measured in terms of a cost function for the risks of death due to vaccination and/or the disease.

To solve the game, we seek a Nash equilibrium strategy. In a population where all individuals play such a strategy, it is impossible for a few individuals to increase their payoffs by switching to a different strategy. Vaccinator cannot be a Nash equilibrium for the reason that an individual who chooses the delayer strategy when population coverage is at 100% reaps the benefits of high population immunity without suffering the risk of vaccine complications. By comparison, delayer can be a Nash equilibrium under certain conditions, such as when the risk of an outbreak is sufficiently low or the risk of death or undesirable effects due to vaccination is sufficiently high. In other situations, it might be best for some individuals to be vaccinated preemptively and for others to delay. To allow for this, let us consider mixed strategies whereby individuals choose the vaccinator strategy with probability $p$ $(0<p<1)$ and the delayer strategy otherwise. If all individuals play the mixed strategy $p$, then a proportion $p = P$ of the population is preemptively vaccinated.

The payoff $E_{vac}$ to an individual choosing the vaccinator strategy is

$$E_{vac} = -d_\nu \tag{11.5}$$

where $d_\nu$ is the probability of adverse effect from the vaccine.

The payoff to an individual choosing the delayer strategy is

$$E_{del}(p) = -r\left[\phi_s(p)d_s + \phi_v(p)d_v\right] \tag{11.6}$$

If

$$E_{vac} > E_{del}(0) \tag{11.7}$$

there is a unique Nash equilibrium $p_{ind}(0 < p_{ind} < 1)$ that can be found by solving for $p_{ind}$ in the equation

$$E_{vac} = E_{del}(p_{ind}) \tag{11.8}$$

If

$$E_{vac} \leq E_{del}(0) \tag{11.9}$$

then the pure delayer strategy ($p_{ind} = 0$) is the unique Nash equilibrium.

Now, if $p$ is the proportion of the population that is preemptively vaccinated in campaigns before outbreaks, we can express the expected cost $C(p)$ due to vaccination and potential outbreaks as

$$C(p) = pd_v + r(1 - p)[(d_s - d_v) \, \phi_s(p) + d_v] \tag{11.10}$$

We then minimize $C(p)$ on the unit interval ($0 \leq p \leq 1$) to determine the group optimum $p_{gr}$, which is the coverage level that would have to be imposed to minimize the total expected number of cases of adverse effects.

# SMALLPOX

The smallpox model (Bauch and Earn, 2004) tracks the time evolution of the densities of individuals who are susceptible ($S$), infected but not-yet infectious ($E$), infectious ($I$), removed (dead/immune) due to smallpox infection ($D$), and removed (dead/immune) due to vaccination ($V$) (the *SEIDV* model). The model equations are:

$$\frac{dS}{dt} = -\beta SI - f(S, \, t)$$

$$\frac{dE}{dt} = \beta SI - \sigma E$$

$$\frac{dI}{dt} = \sigma E - \gamma I \tag{11.11}$$

$$\frac{dD}{dt} = \gamma I$$

$$\frac{dV}{dt} = f(S, \, t)$$

where $\beta$ is the mean transmission rate, $1/\sigma$ is the mean latent period, and $1/\gamma$ is the mean infectious period (here defined as the time between the onset of infectiousness and the isolation of the symptomatic patient). The effect of postattack intervention by health authorities is expressed through the function $f(S, t)$, which is the rate of mass vaccination of susceptible

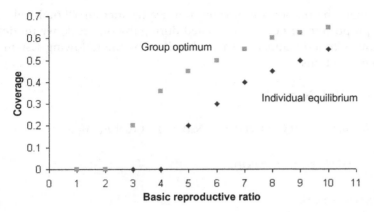

**FIGURE 11.1**

Results of the game for smallpox. Optimum vaccine coverage as a function of the Basic Reproduction Ratio. *From Bauch, C.T., Earn, D.J.D., 2004. Vaccination and the theory of games. Proc. Natl. Acad. Sci. USA 101(36), 13391–13394.*

individuals. We assume that mass vaccination is initiated $t_{res}$ days after the initial smallpox exposure (the "response time") and that susceptible individuals are vaccinated at a constant rate $v$ until all have been vaccinated; hence

$$f(S, t) = \begin{cases} 0 & 0 < t < t_{res} \\ v & t > t_{res} \text{ and } S > 0 \\ 0 & t > t_{res} \text{ and } S = 0 \end{cases} \qquad (11.12)$$

To estimate the initial attack size, we assume that the attack takes the form of aerosolized dispersal in, for instance, a building or airport. Therefore the initial number of individuals exposed to the virus $\alpha$ is large, but a proportion $p$ are immune due to preemptive vaccination. As a result, the initial number of infected individuals is $(1 - p)\alpha$.

The result by Bauch and Earn (2004) is summarized in Fig. 11.1 in which the optimum coverage is displayed as a function of the Basic Reproduction Ratio.

As can be seen, the group optimum vaccine coverage is always greater than the individual equilibrium.

## YELLOW FEVER

The YF game (Massad et al., 2005a,b) proposed has the structure similar to that of the one proposed for smallpox, i.e., a proportion $p$ of susceptibles is preemptively vaccinated and its complement $(1 - p)$ decides to wait until an outbreak is established.

To calculate the proportion of cases and deaths after an YF outbreak, as well as the proportion of people vaccinated during the outbreak, we modeled the epidemic with a dynamical system, described by the following system of differential equations:

$$\frac{dM_s}{dt} = -ca\, M_s H_i / N_h + \mu_M M_s$$

$$\frac{dM_i}{dt} = ca \exp(-\mu_M \tau) M_s(t-\tau) H_i(t-\tau)/N_h(t-\tau) - (\alpha_M + \mu_M) M_s$$

$$\frac{dH_s'}{dt} = -ba M_i H_s'/N_h - (\nu_h + \mu_h) H_s'$$

$$\frac{dH_s''}{dt} = -(\mu_\nu + \mu_h) H_s'' \qquad\qquad (11.13)$$

$$\frac{dH_i}{dt} = ba M_i H_s'/N_h - (\gamma_h + \mu_h + \alpha_h) H_i$$

$$\frac{dH_\nu}{dt} = \nu_h H_s' - (\mu_h + \mu_\nu) H_\nu$$

$$\frac{dH_r}{dt} = \gamma_h H_i - \mu_h H_r$$

Let us now briefly describe some features of the system. Consider first the mosquito population, described by the first two equations of system (11.13). When a susceptible (without the infection) mosquito bites an infected person, it may become (with a certain probability) infected. If it survives for a period of time $\tau$ (the extrinsic incubation period), it became infective, i.e., if it bites a human it may transmit (with a certain probability) the infection. We are not interested in infected but not-yet infective mosquitoes, but rather we consider only susceptible mosquitoes, $M_s$, and infective mosquitoes, $M_i$. The first two equations of the system (11.13) describe the dynamics of those two populations. Consider the first equation. We call $a$ the biting rate of the mosquitoes. So $aM_s$ is the number of bites the susceptible mosquitoes inflict per unit of time in humans. Of those, only a proportion $H_i/N_h$ will be on infected humans, and of those, only a proportion $c$ will result in infected mosquitoes. Susceptible mosquitoes are assumed to die with a rate $\mu_M$. The first term of the second equation describes the number of mosquitoes that became infected $\tau$ units of time earlier, survived a time interval $\tau$ and now became infective. The infective mosquitoes are assumed to die with a rate $\alpha_M + \mu_M$.

Let us now consider the human population. Humans are divided into those who were preemptively vaccinated, denoted $H''_s$, and those who did not receive the vaccine and are, therefore, truly susceptible, denoted $H'_s$. The latter acquired the infection from infective mosquitoes through the bites $aM_i H'_s/N_h$, a fraction of which, $b$, are actually infective. They may be vaccinated during an outbreak, with a rate $\nu_h$, or die by natural causes, with a rate $\mu_h$. The individuals preemptively vaccinated, $H''_s$, die with rates $\mu_v$ (by the effect of the vaccine) and $\mu_h$, the natural mortality rate of humans. Once infected, $H_i$, individuals can either recover from the infection, with rate $\gamma_h$, or die with rates $\alpha_h$ (the mortality rate of YF) or $\mu_h$, the natural mortality rate of humans. Individuals vaccinated during the outbreaks, $H_v$, can die by natural causes, or by the vaccine, with rates $\mu_h$, or $\mu_v$, respectively. Those recovered from the infection, $H_r$, die only by natural causes.

From system (11.13), it is possible to calculate the threshold for the establishment of an epidemic, resulting in

$$T = \frac{N_m}{N_h} \frac{a^2 bce^{-\mu_m \tau}}{(\gamma + \mu_h + \alpha_h)(\mu_m + \alpha_m)} \tag{11.14}$$

Expression (11.14) is the basic reproduction number as defined by Macdonald (1952). Its numerical value can also be estimated through the analysis of the doubling time of the initial phase of an epidemic (Anderson and May, 1991).

The values of $T$ of dengue were calculated for 17 cities of the State of São Paulo, through the doubling time of the initial phase of the epidemic, as described by Anderson and May (1991) and Massad et al. (2001, 2003). Then, we calculate $T$ of YF. Next, we calculated the size of the mosquito population for each city analyzed, by the relation:

$$N_m = \frac{N_h}{T_{yf}} \frac{(\gamma + \mu_h + \alpha_h)(\mu_m + \alpha_m)}{a^2 bce^{-\mu_m \tau}} \tag{11.15}$$

since $N_h$ is known and the other parameters were assumed to be the same for all the cities analyzed because they are all in the same microclimatic region of the State of São Paulo.

System (11.9) was numerically simulated to obtain the quantities necessary to estimate the optimum proportion to vaccinate, $p_{gr}$, that minimizes the total number of deaths:

$$\pi_{yf}(p) = \frac{\int_0^\infty baM_i H'_s/N_h dt}{(1-p)N_h}$$

$$d_{yf}(p) = \frac{\int_0^\infty (\mu_h + \alpha_h)H_i dt}{(1-p)N_h}$$

$$\pi_v(p) = \frac{\int_0^\infty \nu_h H'_s dt}{(1-p)N_h} \tag{11.16}$$

$$d'_v(p) = \frac{\int_0^\infty \mu_v H_v dt}{(1-p)N_h}$$

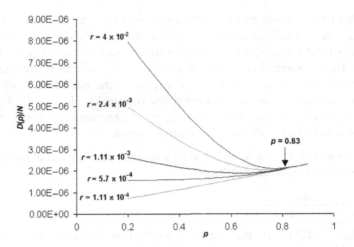

**FIGURE 11.2**

Optimum vaccination coverage that minimizes the total mortality. Different lines denote different risks of outbreaks. *From Massad, E., Coutinho, F.A.B., Burattini, M.N., Lopez, L.F., Struchiner, C.J., 2005b. Yellow fever vaccination: how much is enough? Vaccine 23, 3908–3914.*

where $p$ is the proportion of the population preemptively vaccinated, as described in the main text.

The main result can be summarized in Fig. 11.2. It can be seen that for low risk of outbreaks the optimum equilibrium is no vaccination. As the risk increases, however, there is a minimum in the mortality due to both vaccination and natural mortality. Note also that after a critical level, vaccination always worsens the death toll.

## POLIOMYELITIS

The poliomyelitis model considers the same game structure as the other previous models, namely, a fraction $p$ of the population is preemptively vaccinated in campaigns and a complement $(1 - p)$ is vaccinated only in the course of an outbreak.

The model in its generalized form considers also two competing polio vaccines, the inactivated vaccine, also known as Salk after Jonas Salt its inventor, and an attenuated vaccine, also known as Sabin, after Albert Sabin, its inventor (Plotkin and Plotkin, 1994). However, in spite of presenting the full model we consider here only the case of attenuated Sabin vaccine because of its reversal to virulent strain phenomenon, which is causing a major concern among public health authorities of developed countries.

The full model is described by the following system of equations:

$$\frac{dS}{dt} = \Lambda - pS\beta_1 E_{NI}I - qS\beta_2 E_{VAC}V_{SAB}$$

$$- b\nu_{SAB} - (1-b)\nu_{SALK}S$$
$$+ \gamma R + \tau_{SAB}V_{SAB} + \tau_{SALK}V_{SALK} - \mu S$$

$$\frac{dI}{dt} = pS\beta_1 E_{NI}I + \sigma V_{SAB} - (\delta + \mu + \alpha)I$$

$$\frac{dV_{SAB}}{dt} = qS\beta_2 E_{VAC}V_{SAB}S + b\nu_{SAB}S \qquad (11.17)$$

$$- (\tau_{SAB} + \mu + \sigma)V_{SAB}$$

$$\frac{dV_{SALK}}{dt} = (1-b)\nu_{SALK}S - (\tau_{SALK} + \mu)V_{SALK}$$

$$\frac{dR}{dt} = \delta I - (\gamma + \mu)R$$

where $S$ stands for susceptible children, $I$ for infected, $V_{SAB}$ are the children vaccinated with the attenuated polio vaccine, $V_{SALK}$ are the children vaccinated with the death virus vaccine, and $R$ are the children who recovered from the infection and who will eventually lost the immune protection conferred by natural infection.

The parameters that determine the system dynamics are described in Table 11.1.

**Table 11.1** Biological Meaning of the Parameters/Variables of Eq. (11.17)

| Parameters/Variables | Biological Meaning |
| --- | --- |
| $\Lambda$ | Number of children born per year |
| $\beta_1$ | Contact rate with the wild virus contaminated environment—$E_{NI}$ |
| $\beta_2$ | Contact rate with the Sabin virus contaminated environment—$E_{VAC}$ |
| $p$ | Probability of being infected with environment—$E_{NI}$ |
| $q$ | Probability of being infected with environment—$E_{VAC}$ |
| $b$ | Ratio Sabin/Salk |
| $\nu_{SAB}$ | Sabin vaccination rate |
| $\nu_{SALK}$ | Salk vaccination rate |
| $\tau_{SAB}$ | Loss of immunity of the Sabin vaccine |
| $\tau_{SALK}$ | Loss of immunity of the Salk vaccine |
| $\sigma$ | Reversion to virulence of the Sabin vaccine |
| $\alpha$ | Additional mortality rate of poliovirus |
| $\delta$ | Rate of development of flaccid paralysis |
| $\mu$ | Natural mortality rate of the population |
| $\gamma$ | Recovery rate of paralysis |

This model allowed us to test another game, which is central in the discussion of polio vaccines, namely, the substitution of the attenuated virus Sabin by the death virus Salk vaccine. In countries where the wild virus circulation is reduced to close to zero, the risk of vaccinal paralysis due to the Sabin vaccine, the substitution of this vaccine by the Salk one is already in course. The game of Sabin versus Salk vaccine was simulated to estimate the Nash equilibrium that would minimize the number of flaccid paralysis both by the wild strain of the poliovirus and the attenuated vaccination Sabin strain.

The preliminary results point to the Nash equilibrium with zero proportion of children being vaccinated with the Sabin vaccine and all children being vaccinated with the Salk vaccine. This was obtained even considering a small risk of the wild-type outbreak and with an efficacy attributed to the Sabin twice as much as that of the Salk. Fig. 11.3 summarizes the results.

It can be seen that the number of flaccid paralysis at equilibrium is minimized by the pure strategy vaccination with Salk only.

The interests of individuals versus the interests of the community; this is what Global Health is all about from the Bioethics perspective. It is also a recipe for conflict. In some situations, however, it is the interest of more powerful communities versus the interest of the powerless ones. Needless to say which wins...

When H1N1 threatened the world of a huge pandemic in 2009, the risk of a disaster in the same magnitude of the 1918 great flu pandemic was supported by the existence of about 5 billion susceptible peoples around the

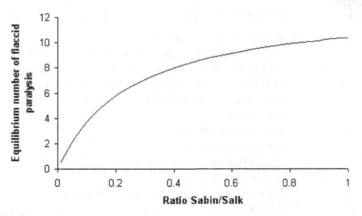

**FIGURE 11.3**

Equilibrium number of flaccid paralysis as a function of the ratio Sabin/Salk vaccines. Note that the number of paralysis is minimized by the substitution of the live by the attenuated vaccine.

world. The world capacity of H1N1 vaccine production was set in the upper bound of 2.8 billion doses for the following 4 years. These facts made evidence an unavoidable conflict; there would not be enough vaccine doses for every susceptible in need. It should be expected that the world producers of the vaccine would keep the dose to their citizens, leaving billions of people unprotected, which is ethically unacceptable. The use of mathematical and computer models, in the same fashion as the one detailed in this chapter, allowed the calculation that vaccinating between 30% and 40% of susceptibles would be enough to mitigate the effects of the pandemic (Massad, unpublished results). Hence the vaccine surplus could be fairly distributed to the nonproducer countries.

The aim of this chapter was to demonstrate how the proper use of information technology can help in the solution of many public health ethical conflicts. The examples detailed above were intended only to illustrate the potential use of the proposed technology in optimizing in an ethically acceptable way, the fair distribution of finite resources.

This chapter examined, so far, what is probably the most common conflict of interest in vaccination (but also in the distribution of any public health intervention in a context of finite resources), namely the interest of individuals (not to be vaccinated provided that everyone else is vaccinated) against the interest of the community (everybody should be vaccinated). This conflict arises because of the well-defined potential risks of vaccines causing serious and sometimes even fatal illness in the recipient of the vaccine or the spreading live-viruses vaccine to immune compromised patients at high risk of developing serious complications. More specifically, we dealt with the application of mathematical models to minimize the conflict of interest in vaccination. Moreover, this chapter examined the moral responsibility of decision makers, the Macroethics issues of controlling infectious diseases.

The approach presented in this chapter shows that it is possible to minimize the conflicts between the individual and the social interests involved in micro and macroethical issues in public health, provided the right tools are applied. In addition, the models presented are probably the only available alternatives to decision-making in many public health dilemmas such as the example of the substitution of live attenuated polio vaccine by the inactivated vaccine. It would be ethically impossible to carry out a case—control study to estimate the strategy that would minimize the cases of flaccid paralysis. Hence the obvious choices are the models of the type presented in this work.

As mentioned earlier, vaccination policies for some infectious diseases present individuals with a conflict: if a sufficient proportion of the population is already immune, either naturally or by vaccination, then even the slightest inconvenience associated with vaccination will outweigh the risk from infection. As a result, individual self-interest might preclude complete eradication of a vaccine-preventable disease. In this chapter it is shown that a formal game theoretical analysis of this problem leads to new insights that help to explain human decision-making with respect to vaccination. In addition, it is showed that the Nash equilibrium, whenever possible to be estimated, provides the optimum strategy in conflict situation. Hence, both in the cases of smallpox and YF, two vaccines with a nonnegligible death rate associated with them, if the risk of outbreaks by the wild virus is too small, it may be better not to vaccinate. In the case of the attenuated OPV with the Sabin strain, the conflict is the one currently in course in developed nations that do not present any case of wild virus paralysis for several years. In those places a substantial number of paralysis cases induced by the attenuated vaccine is causing major concerns among public health authorities. The model here presented tested several mixed strategies with both the Sabin and Salk vaccine combinations and concluded that the Nash equilibrium that minimizes the number of flaccid paralysis is obtained by the pure Salk vaccination.

Bioethics should take into account conflicts between individuals and the community such as the conflict arisen by vaccination against infectious diseases. More important, however, is the conflict of interest between unequal communities or countries for the distribution of intervention strategies in a context of finite resources. In the presence of such an efficacious tool as mathematical and computer models, which have demonstrated their potential as optimizer of public health strategies, it would be unethical not to apply them as an auxiliary technique in decision-making to mitigate potential conflicts.

## SPILLOVER: A REAL TRANSBORDER PROBLEM

The term Spillover is normally used to describe zoonotic infections that cross the animal—human hosts barrier, establishing themselves as emergent infections (Quammen, 2012). Examples abound and it is estimated that all new infections that emerged in the last decades had an animal origin (The Lancet Editorial, 2004). In this chapter, however, spillover represents outbreaks of emerging or reemerging infections in one country or region of the world that threaten other countries or regions, spreading by human (or in some cases migrating birds or mosquitoes) movements. The recent outbreak of Ebola virus in West Africa and the current outbreak of Zika virus in Latin America exemplify, in a dramatic way the ethical issues related to the threat of emerging infections to unaffected regions of the world. The risk of spillover and the power inequalities to quell it are discussed, as well as the perspective of information technologies in providing the necessary tools to solve the ethical dilemmas posed by such threats.

## IMPORTATION AND EXPORTATION OF INFECTIONS

The spillover of an emergent or reemergent infection may occur in two ways: importation or exportation of cases. Importation refers to the case when a traveler from a disease-free country or region visits an endemic or epidemic area and gets infected (Lopez et al., 2016). He/she can then carry the infection back home, where the infection may eventually establish itself, provided the basic reproduction ratio of the infection is greater than one. Exportation, in turn, refers to the case when an inhabitant of an endemic or epidemic country or region travels infected to a disease-free area. There, again, the infection may eventually establish itself, provided the basic reproduction ratio of the infection is greater than one. Of course a combination of both importation and exportation may occur when, say, in a plane flying from an endemic or epidemic region of country both infected visitors and/or local inhabitants travel to a disease-free country or region. The important difference, however, is that visitors always travel back, either in the overly diseased (if not too ill to be in the hospital), whereas local inhabitants travel only in the incubating condition. Infections with long incubation periods, therefore, are more prone to be exported from endemic/epidemic regions/countries to disease-free regions/countries. In this section the cases of exportation of infections will be exemplified by the case of Ebola virus from West Africa (Lopez et al., 2016) and Zika virus from Brazil (Massad et al., 2016).

## THE CASE OF EBOLA EXPORTATION FROM WEST AFRICA

The case of disease exportation is simpler, from the modeling point of view, than the infection importation. The basic difference is that in the latter, visitors return infective to their home country. Their number can be calculated as follows (Lopez et al., 2016):

$$I(t) = \sigma' S(0) e^{-(\gamma+\mu+\alpha)t} \int_0^t \left[ e^{-\sigma' t} \int_0^{t'} \lambda(t) e^{-\int_0^\tau \lambda(s)ds - (\sigma-\sigma'-\gamma-\alpha)\tau} d\tau \right] dt' \qquad (11.18)$$

where the variables and parameters are detailed in Table 11.2.

In the exportation case, in contrast, individuals depart from the endemic home country in a latent state, i.e., already infected but still asymptomatic. This latter assumption is based on the conjecture that infective and symptomatic individuals do not travel. Their disease will manifest itself either during the voyage or after arrival to the disease-free country. Another important

**Table 11.2** Biological Meaning of the Parameters/Variables of Eq. (11.18)

| Parameters/Variables | Biological Meaning |
|---|---|
| $I(t)$ | Travelers that return infected from endemic areas/countries |
| $S(0)$ | Susceptible travelers arriving in the endemic regions at time $t = 0$ |
| $\sigma'$ | Infected visitors' rate of returning back to their regions/countries |
| $\sigma$ | Noninfected visitors' rate of returning back to their regions/countries |
| $\gamma$ | Recovering rate of the disease |
| $\mu$ | Natural mortality rate of visitors |
| $\alpha$ | Disease-induced mortality rate of visitors |
| $\lambda(t)$ | Force of infections (incidence density) of the disease |
| $\tau$ | Moment at which visitors get the infection |

difference between the disease importation and exportation models is that in the former, the key parameter was the force of infection of the disease in the visited endemic-country, whereas in the latter the key parameter is the latency duration of the disease. In terms of the modeling, in the disease importation case, latency is not too important and the model considers only susceptible, infected and removed individuals. Hence the number of new cases per unit of time corresponds to the infection incidence, denoted $\lambda(t)S(t)$. In the case of disease exportation, however, latency is important because it is assumed that infected and symptomatic individuals are either so sick that they do not manage to travel or are prohibited of doing so. In a SEIR (Susceptible-Exposed-Infective-Recovered) type of model (Anderson and May, 1991), we can calculate the probability that one infected but still not infectious individual is leaving and endemic region/country at time between $t$ and $t + dt$, $p_E(t)$, according to:

$$p_E(t) = p_E(0)e^{-(\mu+\alpha+\delta)t} + \frac{1}{N}\int_0^t \frac{\beta S(t')(I')}{N}e^{(\mu+\alpha+\delta)t'}dt' \qquad (11.19)$$

where the variables and parameters are detailed in Table 11.3.

Multiplying the number of visitors to a given disease-free country by the prevalence of latent individuals, $p_E(t)$, and by the prevalence of infected individuals, $p_I(t)$, generates the number of infected visitors or of infections exportation.

To obtain this prevalence, the force of infection of the disease in this endemic region is a necessary input variable. Unfortunately the best information normally available is the notification rate of infectious individuals, provided by disease surveillance systems. The next section shows one possible way out to circumvent this limitation and illustrates an actual case of a recent

**Table 11.3** Biological Meaning of the Parameters/Variables of Eq. (11.19)

| Parameters/ Variables | Biological Meaning |
| --- | --- |
| $p_E(0)$ | Probability of one latent individual at the moment immediately before traveling |
| $S(t)$ | Susceptible travelers in the endemic regions at the traveling back time |
| $I(t)$ | Infected travelers in the endemic regions at the traveling back time |
| $\delta$ | Inverse of the disease incubation period |
| $\beta$ | Potentially infective contact rate between susceptible and infected individuals |
| $\mu$ | Natural mortality rate of visitors |
| $\alpha$ | Disease-induced mortality rate of visitors |
| $N$ | Total number of individuals at risk |

disease outbreak that is spreading itself for previously disease-free countries, namely, the Ebola epidemic in West Africa.

As of December 17, 2014, the Ebola outbreak had already affected 18,603 people in some countries of West Africa, with 6915 confirmed fatalities (WHO, 2014). A few cases have already reached previously unaffected countries, like the United States and Spain. A few cases have already reached previously unaffected countries outside of the African continent, including the United States and Spain. Although some West African countries with infections early this year have already been declared free of disease, the Ebola epidemic continues to wreak havoc in Liberia, Guinea, and Sierra Leone.

The current Ebola outbreak in West Africa is used to illustrate the exportation model. The theory is applied in Liberia, one of the worst affected countries in terms of number of cases and deaths.

The reported number of cases, $\delta E(t)$, can be fitted to a continuous function, e.g.:

$$\delta E(t) = \kappa_1 \text{sech}^2(\kappa_2 t + \kappa_3) \tag{11.20}$$

where $\kappa_i$, $i = 1, 2, 3$ are fitting parameters and $E(t)$ is the number of latent individuals. If the duration of Ebola incubation $1/\delta$ is known, it is possible to estimate the probability of at least one individual is incubating the virus when traveling abroad.

The result for the case of Ebola in Liberia is shown in Fig. 11.4.

In an example cohort of travelers that depart from Liberia at Week 15, the relative number of latent individuals carrying the Ebola virus is of 0.3 individuals per 1000 travelers, which is pretty much in accord to the imported number of Ebola reported.

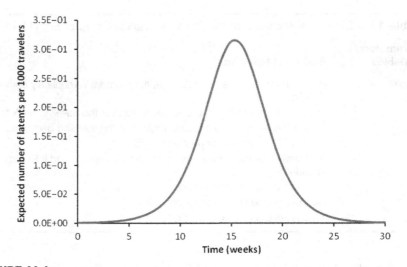

**FIGURE 11.4**

Expected number of incubating individuals per 100,000 travelers. *From Lopez, L.F., Amaku, M., Coutinho, F.A., Quam, M., Burattini, M.N., et al., 2016. Modeling importations and exportations of infectious diseases via travelers. Bull. Math. Biol. 78, 185–209.*

## THE CASE OF ZIKA VIRUS EXPORTATION FROM BRAZIL

Zika virus (ZIKV) infections were first reported in Brazil in May 2015, and by December 2015, 500,000—1,500,000 of ZIKV infections were estimated (ECDC, 2015, 2016). By October 2015, increasing number of microcephaly cases and other neonatal malformations were thought to be associated with ZIKV infections (Lancet Editorial, 2016). On February 1, 2016, the clusters of microcephaly and Guillain-Barré Syndrome (GBS) cases in likely association with ZIKV infections were declared a public health emergency of international concern (Heymann et al., 2016). Given the interconnectivity of Brazil with the remainder of the world, ZIKV has the potential to spread rapidly around the world via viremic travelers (Bogoch et al., 2016). The extent of spread depends on the travel volume to destination countries and the endemicity in the exporting country (Quam and Wilder-Smith, 2015; Wilder-Smith et al., 2014; Tatem et al., 2012). Because of the mild clinical manifestations of the disease in the vast majority of cases, ZIKV infections in individual travelers are unlikely to lead to cancelation of flights or disruption of holiday/business plans. Furthermore, 80% of all infections are thought to be asymptomatic. The biggest concern is the spread to areas where suitable mosquito vectors exist and importation could trigger further outbreaks. However, given that sexual transmission of ZIKV has been reported, the spread of ZIKV via viremic travelers to areas without the *Aedes* mosquitoes is equally of concern (Hills et al., 2016). Sexual transmission to nontraveling contacts in Europe could propagate ZIKV infections in Europe, resulting in a potential upsurge of GBS cases as a result of imported ZIKV infections and putting pregnant women at risk. Therefore it is important to estimate the potential number of travelers returning to Europe with ZIKV infections.

ZIKV infections remain underdiagnosed and underreported due to the non-specific and mild manifestations and lack of widely available diagnostic assays. Therefore, for the time being any estimates on the epidemiological burden remain crude estimates. We based our calculations on the published estimate of 500,000 to 1,500,000 infections (both symptomatic and asymptomatic) for the year 2015 in Brazil (ECDC, 2016). Reliance on reported events of importation will only underestimate the true importation risk as most imported cases will not be detected and reported, unless the clinical manifestations are more severe. In the absence of reliable surveillance data, mathematical modeling is necessary to estimate the number of importations of ZIKV from Brazil into Europe.

A previously developed mathematical model on exportations (Lopez et al., 2016) was used to estimate the number of Zika virus importations into Europe. This model takes into account the travel volume, the probability of being infected at the time of travel, the population size of Brazil, and the estimated incidence of ZIKV infections (estimated numbers over population size). The model was previously developed to estimate the risk of dengue acquisition in international travelers, (Massad et al., 2013, 2014; Massad and Wilder-Smith, 2009) and has also been applied to estimate poliovirus importations (Wilder-Smith et al., 2015).

The number of air passenger numbers from Brazilian airports to each of the European countries was obtained from the International Air Transport Association for the year 2012, per month and per destination country in Europe; we only had access to the year 2012 flight data, in 2015 or 2016 the travel pattern of outgoing flights will not have changed significantly.

The force of infection, $\lambda(t)$, was estimated from the assumption that there had been 0.5−1.5 million ZIKV infections in Brazil. In addition, we assumed that the seasonal distribution of cases followed the same as for dengue, given that both viral infections share the same *Aedes* vectors, and initial observations have claimed that Zika seems to follow the path of dengue (Musso et al., 2015a). As populations of *A. aegypti* and *A. albopictus* are climate sensitive and display a seasonal pattern in Brazil (Lowe et al., 2016; Cavalcanti et al., 2010; Siqueira et al., 2005), Zika virus infections are likely to exhibit the same seasonal pattern as dengue in Brazil.

A continuous function was fitted to the time distribution of notified cases from which we estimate the force of infection $\lambda(t)$. The product of the force of infection by the fraction of susceptible individuals is the number of reported cases.

The individual risk of acquiring the infection from the Zika-infected mosquitoes, $Risk(t)$, is given by

$$Risk(t) = 1 - \exp\left[-\int_{t_1}^{t_2} \lambda(t)dt\right] \tag{11.21}$$

where, again, $\lambda(t)$ is the force of infection or incidence density rate, $t_1$ is, in the case of travelers, the moment they arrive at the endemic area and $t_2$ is the moment they depart. Note that the concept of risk expressed in Eq. (11.1) means the risk for travelers that remain in the Zika endemic area for the period between $t_1$ and $t_2$. For locals, $t_2 - t_1$ is the time interval considered for the risk calculation (e.g. the month-by-month risk calculation).

The risk varies with time, and this risk is highest in the months with the highest density of infected mosquitoes, attaining its maximum by the month of April. This would also fit with the observation of the onset of excess microcephaly cases in October 2015 (6−9 months after the high season of January to April).

As the function $Risk(t)$ represents the individual risk of acquiring the infection, we can use it as the probability that one passenger flying from a Brazilian airport is infected with the Zika virus. By multiplying the individual probability of being infected by the number of passengers leaving Brazilian airports, the total number of expected infections that are flying to European countries are calculated.

The model applies to individuals from Brazil traveling to Europe or travelers having visited Brazil and now returning to Europe.

Fig. 11.5 shows the results of the expected number of ZIKV exported to European countries from Brazil, based on an estimated lower bound of 500,000 and upper bound of 1, 500, 000 ZIKV infections, respectively, assuming that these ZIKV infections exhibit the same seasonal pattern as dengue infections. In total, the model estimated between 508 and 1778 imported cases, respectively, into all European countries, with the highest numbers being in France, Portugal, and Italy. Of these, 80% would likely be asymptomatic, hence we would expect between 116 and 355 symptomatic ZIKV infections.

The model's estimates are consistent with those reported by the European Center for Disease Control. As of March 3, 2016, ECDC had recorded 209 imported cases into 16 European countries, of which 81 were into France and 32 into Spain (ECDC, 2016) Geographical distribution of Zika virus has

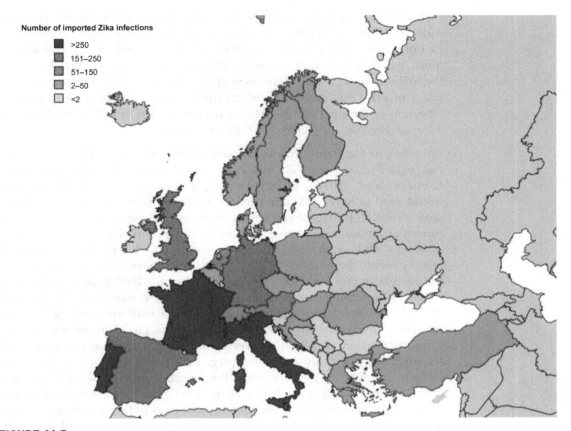

**FIGURE 11.5**

Estimated importations of Zika virus infections via travelers from Brazil to Europe in the year 2015, based on a high estimate of 1,500,000 Zika virus infections in Brazil. *From Massad, E., Tan, S.H., Kahn, K., Wilder-Smith, A., 2016. Estimated Zika virus importations to Europe by travellers from Brazil. Glob. Health Action 9, 31669 - <http://dx.doi.org/10.3402/gha.v9.31669>.*

steadily broadened since the virus was first detected in Brazil in 2015. By March 2016, Zika transmission has been reported in 28 countries/territories (WHO, 2016), hence the exportation risk will be even higher than we reported. However, we were not able to calculate such a risk for the other countries as incidence data for those countries have not yet been published. Given that Brazil so far has been the country most affected with the highest absolute numbers of estimated ZIKV infections, it is justified to focus our model on Brazil as exporting country only, until more data are available from other Latin American countries.

A limitation of our study is that the underlying assumption of our model is the equal distribution of cases throughout the country, and the equal probability of traveling throughout the Brazilian population. However, in early 2015, the geographic concentration of most cases were in the North East of Brazil—but by late 2015 and early 2016, the distribution was already much wider spread with all major cities in Brazil affected (ECDC, 2016; Heukelbach et al., 2016; Hennessey et al., 2016). Hence our modeled estimates of ZIKV exportations based on travel volume will be a more accurate reflection of the situation in 2016, assuming that the year 2016 will also see between 500,000 and 1,500,000 ZIKV infections.

According to the French Polynesian case—control study on ZIKV related GBS, one would expect 24 GBS cases out of 100,000 ZIKV infections (Cao-Lormeau et al., 2016). In other words, if these estimates hold true, one would need to have 5000 imported ZIKV infections to see one case of ZIKV-associated GBS in returning travelers from ZIKV-affected countries. Given the current exportation numbers estimated to be no more than 1800, the probability of a ZIKV-associated GBS case in Europe in 2015 or 2016 is relatively low. However the number of ZIKV affected countries within Latin America, the Caribbean and beyond is rising, and hence the likelihood of substantial number of returning travelers presenting with GBS is also likely. The true risk of ZIKV-related infections that can lead to CNS malformations and microcephaly in pregnant women is currently unknown, especially for sexual transmission (Brasil et al., 2016). However, potentially every single viremic male returning traveler could infect his pregnant or nonpregnant partner, especially in the first 2—4 weeks after ZIKV infection (Foy et al., 2011; Musso et al., 2015b; Venturi et al., 2016). Hence the Center for Disease Control and travel medicine providers have advised for precautions (abstinence or condoms) to be taken for men returning home from ZIKV-affected countries, particularly in the first few weeks after return (Vouga et al., 2016; Goorhuis et al., 2016).

An additional cause of concern is the risk of ZIKV establishing itself in European regions where the presence of *A. albopictus* is endemic, in particular for Mediterranean countries recording increasingly hotter summers (Helmerson et al., 2016).

The model identified high-risk countries in Europe, and policy-makers and clinicians need to be aware of such data. Furthermore, our models can be applied by individual countries or by continents alike and used as an additional tool to estimate the risk of importation based on the main contributing factors such as travel volume and the evolving Zika virus endemicity in exporting countries. The model help policy-makers estimate the extent of importations to prepare for the scale up of laboratory diagnostic assays, estimate the occurrence of GBS, potential sexual transmission, and infants with congenital Zika virus syndrome.

## CODA

At the time of writing this chapter a urban YF outbreak raged throughout Angola, which has infected more than 355 people and killed 159—the first epidemic of the disease to hit the country in 30 years (WHO, 2016a). The epidemic has already spilt over to the Democratic Republic of Congo (DRC), Mauritania, Kenya, and China, the first time YF is reported in Asia (Woodall and Yuill, 2016). As of March 24, 2016, WHO and partners have vaccinated 5.7 million people in Luanda, the Angolan capital, against YF using vaccines from the International Coordination Group emergency stockpile (WHO, 2016b). The vaccination campaign covered almost the entire population of Luanda but experts estimate that more than 10 million additional doses of the vaccine will be necessary to cover the rest of the country (Woodall and Yuill, 2016). Kinshasa, the capital of DRC alone, has a population of more than 11 million people and it is an important hub of air flights to the rest of Africa and the world. Although YF vaccine production grew from less than 30 million yearly doses to more than 83 million doses today (Chan, 2016), this is far from the necessary to curb this imminent thread to *Aedes*-infested cities of the world. The solutions proposed include diluting the original doses up until 1-to-10 such that the 5 doses vial could vaccinate 50 people (Woodall and Yuill, 2016). These desperate measures, however, are prone to generate discredit among the affected populations. The methods proposed in this chapter to optimize vaccination strategies could be of immense help and as demonstrated, there is no need to vaccinate the entire populations of the countries at risk. It is a moral responsibility of the public health decision-makers involved to attempt maximizing the limited resources available to control outbreaks of these infections.

I use this tragic example of an urgent macroethical problem such as the resurgence of urban YF to finish this chapter with a summary of the ethical and transborder issues. I quote Woodall and Yuill (2016): "New York city just announced a 3-year, $21 million plan to protect New Yorkers from the Zika virus. Imagine how much of that could have been saved by putting just one-tenth of the money into the prevention of Ebola and Zika at the source". In addition to the moral responsibility of those who have to those who do not have (Macroethics), it should be of the interest of the former to curb spill-over of emergent and reemergent infections at the source countries. Among other issues, it is cheaper...

## References

Anderson, R.M., May, R.M., 1991. Infectious Diseases of Humans: Dynamics and Control. Oxford University Press, Oxford.

Bassingthwaighte, J.B., 2003. The macro-ethics of genomics to health: the physiome project. C. R. Biol. 2003 326 (10−11), 1105−1110.

Battin, M.P., Francis, L.P., Jacobson, J.A., Smith, C.B., 2009. The Patient as Victim and Vector: Ethics and Infectious Diseases. Oxford University Press, Oxford.

Bauch, C.T., Earn, D.J.D., 2004. Vaccination and the theory of games. Proc. Natl. Acad. Sci. USA 101 (36), 13391–13394.

Bauch, C.T., Galvani, A.P., Earn, D.J.D., 2003. Group interest versus self-interest in smallpox vaccination policy. Proc. Natl. Acad. Sci. USA 100, 10564–10567.

Beauchamp, T., Childress, J., 2001. Principles of Biomedical Ethics. Oxford University Press, Oxford, NY.

Blum, J.D., Talib, N., 2006. Balancing individual rights versus collective good in public health enforcement. Med. Law 25 (2), 273–281.

Bogoch, I.I., Brady, O.J., Kraemer, M.U., German, M., Creatore, M.I., et al., 2016. Anticipating the international spread of Zika virus from Brazil. Lancet 387, 335–336.

Bossak, B.H., Welford, M.R., 2009. Did medieval trade activity and a viral etiology control the spatial extent and seasonal distribution of Black Death mortality? Med. Hypotheses 72, 749–752.

Brasil, P., Pereira Jr., J.P., Raja Gabaglia, C., Damasceno, L., Wakimoto, M., et al., 2016. Zika virus infection in pregnant women in Rio de Janeiro—preliminary report. N. Engl. J. Med.

Briggle, A., Mitcham, C., 2012. Ethics and Science: An Introduction. Cambridge University Press, Cambridge.

Caley, P., Becker, N.G., Philp, D.J., 2007. The waiting time for inter-country spread of pandemic influenza. PLoS One 2, e143.

Cao-Lormeau, V.M., Blake, A., Mons, S., Lastere, S., Roche, C., et al., 2016. Guillain-Barre syndrome outbreak associated with Zika virus infection in french polynesia: a case-control study. Lancet 387, 1531–1539.

Cavalcanti, L.P., Coelho, I.C., Vilar, D.C., Holanda, S.G., Escossia, K.N., et al., 2010. Clinical and epidemiological characterization of dengue hemorrhagic fever cases in northeastern, Brazil. Rev. Soc. Bras. Med. Trop. 43, 355–358.

CDC, 2002. Adverse events associated with 17D-derived yellow fever vaccination—United States, 2000–2002. MMWR Morb. Mortal. Wkly. Rep. 51 (44), 989–993.

CDC, 2003. Health topics: poliomyelitis. <http://www.cdc.gov/health> (accessed 20.12.03).

Chan, M., 2016. Yellow fever: the resurgence of a forgotten disease. Lancet 387, 2165–2166.

Chan, R.C., Penney, D.J., Little, D., Carter, I.W., Roberts, J.A., Rawlinson, W.D., 2001. Hepatitis and death following vaccination with 17D–204 yellow fever vaccine. Lancet 358, 121–122.

DHEW, 1978. National Commission for the Protection of Human Subjects of Biomedical and Behavioral Research. Department of Health, Education and Welfare.

Dietz, K., Heesterbeek, J.A.P., 2002. Daniel Bernoulli's epidemiological model revisited. Math. Biosci. 180 (1–2), 1–21.

Dixit, A., Skeat, S., 1999. Games of Strategy. W.W. Norton and Company, New York, NY.

ECDC, 2015. Rapid risk assessment: Zika virus epidemic in the Americas: potential association with microcephaly and Guillain-Barré syndrome.

ECDC, 2016. Epidemiological situation of the Zika outbreak.

Fabro, J.A., Jacobson, V., 1970. Massachusetts 25S. Ct. 358 (1905): compulsory vaccination. Conn. Med. 34 (9), 654–655.

Foy, B.D., Kobylinski, K.C., Chilson Foy, J.L., Blitvich, B.J., Travassos da Rosa, A., et al., 2011. Probable non-vector-borne transmission of Zika virus, Colorado, USA. Emerg. Infect. Dis. 17, 880–882.

Gomes, M.F.C., Piontti, A.P., Rossi, L., Chao, D., Longini, I., Halloran, M.E., et al., 2014. Assessing the international spreading risk associated with the 2014 West African Ebola outbreak. PLoS Curr. Outbreaks September 2, 2014. Edition 1. <http://dx.doi.org/10.1371/currents.outbreaks.cd818f63d40e24aef769dda7df9e0da5>.

Goorhuis, A., von Eije, K.J., Douma, R.A., Rijnberg, N., van Vugt, M., et al., 2016. Zika virus and the risk of imported infection in returned travelers: implications for clinical care. Travel Med. Infect. Dis. 14, 13−15.

Halloran, M.E., Longini, I.M., Nizam, A., Yang, Y., 2002. Containing bioterrorist smallpox. Science 298, 1428−1432.

Helmerson, J.L., Quam, M., Wilder-Smith, A., Stenlund, H., Ebi, K., Massad, E., et al., 2016. Climate change and *Aedes* vectors: 21st century projections for dengue transmission in Europe. EBiomedicine. Available online April 2, 2016. <http://dx.doi.org/10.1016/j.ebiom.2016.03.046>.

Henderson, D.A., Moss, B.M., 1999. In: Plotkin, S.A., Orenstein, W.A. (Eds.), Smallpox and vaccinia, in vaccines, third ed. S.W. Saunders Company, Philadelphia, pp. 74−97. (Chapter 6).

Hennessey, M., Fischer, M., Staples, J.E., 2016. Zika virus spreads to new areas—region of the Americas, May 2015−January 2016. MMWR Morb. Mortal. Wkly. Rep. 65, 55−58.

Herkert, J.R., 2004. Microethics, macroethics, and professional engineering societies. National Academy of Engineering, Emerging Technologies and Ethical Issues in Engineering: Papers from a Workshop, October 14−15, 2003 ISBN 978-0-309-09271-5.

Heukelbach, J., Alencar, C.H., Kelvin, A.A., De Oliveira, W.K., Pamplona de Goes Cavalcanti, L., 2016. Zika virus outbreak in Brazil. J. Infect. Dev. Ctries 10, 116−120.

Heymann, D.L., Hodgson, A., Sall, A.A., Freedman, D.O., Staples, J.E., et al., 2016. Zika virus and microcephaly: why is this situation a PHEIC? Lancet 387, 719−721.

Hills, S.L., Russell, K., Hennessey, M., Williams, C., Oster, A.M., et al., 2016. Transmission of Zika virus through sexual contact with travelers to areas of ongoing transmission—continental United States, 2016. MMWR Morb. Mortal. Wkly. Rep. 65, 215−216.

Jonsen, A.R., 1998. The Birth of Bioethics. Oxford University Press, Oxford.

Kaplan, E.H., Craft, D.L., Wein, L.M., 2002. Emergency response to a smallpox attack: the case for mass vaccination. Proc. Natl. Acad. Sci. USA 99, 10935−10940.

Khan, K., Memish, Z.A., Chabbra, A., et al., 2010. Global public health implications of a mass gathering in Mecca, Saudi Arabia during the midst of an influenza pandemic. J. Travel. Med. 17, 75−81.

Lopez, L.F., Amaku, M., Coutinho, F.A., Quam, M., Burattini, M.N., et al., 2016. Modeling importations and exportations of infectious diseases via travelers. Bull. Math. Biol. 78, 185−209.

Lowe, R., Coelho, C.A., Barcellos, C., Carvalho, M.S., Catao Rde, C., et al., 2016. Evaluating probabilistic dengue risk forecasts from a prototype early warning system for Brazil. eLife 5, e11285.

Macdonald, G., 1952. The analysis of equilibrium in malaria. Trop. Dis. Bull 49, 813−828.

Martin, M., Tsai, T.F., Cropp, B., Chang, G.J., Holmes, D., Tseng, J., et al., 2001. Fever and multi-system organ failure associated with 17D−204 yellow fever vaccination: a report of four cases. Lancet 358, 98−104.

Mason, R.O., 1995. Applying ethics to information technologies issues. Commun. ACM 38, 55−57.

Massad, E., Wilder-Smith, A., 2009. Risk estimates of dengue in travelers to dengue endemic areas using mathematical models. J. Travel. Med. 16, 191−193.

Massad, E., et al., 1994. The basic reproduction ratio of HIV among intravenous-drug-users. Math. Biosci 123, 227−247.

Massad, E., Coutinho, F.A.B., Burattini, M.N., Lopez, L.F., 2001. The risk of yellow fever in a dengue infested area. Trans. R. Soc. Trop. Med. 95 (3), 370−374.

Massad, E., Burattini, M.N., Coutinho, F.A.B., Lopez, L.F., 2003. Dengue and the risk of urban yellow fever reintroduction in São Paulo, Brazil. Rev. Saúde Pública 37 (4), 477−484.

Massad, E., Coutinho, F.A.B., Burattini, M.N., Lopez, L.F., 2004. The Eyam plague revisited: did the village isolation change transmission from fleas to pulmonary? Med. Hypotheses 63, 911−915.

Massad, E., Burattini, M.N., Lopez, L.F., Coutinho, F.A.B., 2005a. Forecasting versus projection models in epidemiology: the case of the SARS epidemics. Med. Hypotheses 65, 17−22.

Massad, E., Coutinho, F.A.B., Burattini, M.N., Lopez, L.F., Struchiner, C.J., 2005b. Yellow fever vaccination: how much is enough? Vaccine 23, 3908−3914.

Massad, E., Burattini, M.N., Coutinho, F.A.B., Lopez, L.F., 2007. The 1918 influenza A epidemic in the city of Sao Paulo, Brazil. Med. Hypotheses 68, 442−445.

Massad, E., Burattini, M.N., Coutinho, F.A.B., Struchiner, C.J., 2010. The risk of acquiring the new influenza A(H1N1) for Brazilian travelers to Chile, Argentina and the USA. Mem. Inst. Oswaldo Cruz. 105, 179−183.

Massad, E., Rocklov, J., Wilder-Smith, A., 2013. Dengue infections in non-immune travellers to Thailand. Epidemiol. Infect. 141, 412−417.

Massad, E., Wilder-Smith, A., Ximenes, R., Amaku, M., Lopez, L.F., et al., 2014. Risk of symptomatic dengue for foreign visitors to the 2014 FIFA World Cup in Brazil. Mem. Inst. Oswaldo Cruz 109, 394−397.

Massad, E., Tan, S.H., Kahn, K., Wilder-Smith, A., 2016. Estimated Zika virus importations to Europe by travellers from Brazil. Glob. Health Action, 2016 9, 31669, <http://dx.doi.org/10.3402/gha.v9.31669>.

Melnick, J.L., 1978. Advantages and disadvantages of killed and live poliomyelitis vaccine. Bull. WHO 56, 21−38.

Monath, T.P., 2001. Yellow fever: un update. Lancet Infect. Dis. 1, 11−20.

Musso, D., Cao-Lormeau, V.M., Gubler, D.J., 2015a. Zika virus: following the path of dengue and chikungunya? Lancet 386, 243−244.

Musso, D., Roche, C., Robin, E., Nhan, T., Teissier, A., et al., 2015b. Potential sexual transmission of Zika virus. Emerg. Infect. Dis. 21, 359−361.

Osborne, M.J., Bubistein, A., 1994. A Course in Game Theory. MIT Press, Cambridge, MA.

Pandey, A., Atkins, K.E., Medlock, J., Wenzel, N., Townsend, J.P., Childs, J.E., et al., 2014. Strategies for containing Ebola in West Africa. Science 346, 991. Available from: http://dx.doi.org/10.1126/science.1260612.

Parmet, W.E., Scott, C., Hodge Jr., J.G., Nahmias, D.E., DeMaria Jr., A., Rees, C.M., 2005. Plenary program: Jacobson v. Massachusetts. J. Law Med. Ethics 33 (4 Suppl.), 24−27.

Plotkin, S.L., Plotkin, S.A., 1994. A Short History of Vaccination. In: Plotkin, S.A., Mortimer, E.A. (Eds.), Vaccines. S.W. Saunders Company, Philadelphia (Chapter 1), pp 1−12, 1994.

Poland, J.D., Calisher, C.H., Monath, T.P., Downs, W.G., Murphy, K., 1981. Persistence of neutralizing antibody 30−35 years after immunization with 17D yellow fever vaccine. Bull. World Health Organ. 59 (6), 895−900.

Quam, M.B., Wilder-Smith, A., 2015. Importation index of dengue to determine the most probable origin of importation. J. Travel Med. 22, 72.

Quammen, D., 2012. Spillover: Animal Infections and the Next Human Pandemic. W.W. Norton & Company, New York, NY.

Rinčić, I., Muzur, A., 2012. Fritz Jahr and the Birth of European Bioethics. Pergamena, Zagreb, 2012, p. 141.

Selgelid, M.J., 2005. Ethics and infectious diseases. Bioethics 19 (3), 272−289.

Siqueira Jr., J.B., Martelli, C.M., Coelho, G.E., Simplicio, A.C., Hatch, D.L., 2005. Dengue and dengue hemorrhagic fever, Brazil, 1981−2002. Emerg. Infect. Dis. 11, 48−53.

Stannard, D.E., 1993. Disease, human migration, and history. In: Kipple, K.E. (Ed.), The Cambridge World History of Human Disease. Cambridge University Press, Cambridge, pp. 35−44.

Struchiner, C.J., Luz, P.M., Dourado, I., Sato, H.K., Aguiar, S.G., Ribeiro, J.G.L., et al., 2004. Risk of fatal adverse events associated with 17DD yellow fever vaccine. Epidemiol. Infect. 132 (5), 939−946.

Tatem, A.J., Huang, Z., Das, A., Qi, Q., Roth, J., et al., 2012. Air travel and vector-borne disease movement. Parasitology 139, 1816−1830.

The Lancet Editorial, 2004. Avian influenza: the threat looms. Lancet 363 (9405), 257.

The Lancet Editorial, 2016. Zika Virus: A New Global Threat for 2016. Lancet 387, 96.

Vallero, D.A., 2007. Beyond responsible conduct in research: new pedagogies to address macro-ethics of nanobiotechnologies. J. Long Term Eff. Med. Implants. 2007 17 (1), 1−12.

Valleron, A.J., 2000. Les rôles de la modélisation en épidémiologie (Roles of mathematical modelling in epidemiology). C. R. Acad. Sci. 323 (5), 429−433.

Vasconcelos, P.F.C., The Brazilian Yellow Fever Vaccine Evaluation Group, 2001. Serious adverse events associated with yellow fever 17DD vaccine in Brazil: a report of two cases. Lancet 358, 91−97.

Velimirovic, B., 1987. AIDS as a social phenomenon. Soc. Sci. Med. 25 (6), 541−552.

Venturi, G., Zammarchi, L., Fortuna, C., Remoli, M.E., Benedetti, E., et al., 2016. An autochthonous case of Zika due to possible sexual transmission, Florence, Italy, 2014. Euro. Surveill. 21.

Vouga, M., Musso, D., Van Mieghem, T., Baud, D., 2016. CDC guidelines for pregnant women during the Zika virus outbreak. Lancet 387, 843−844.

WHO, 2003. International travel and health. Vaccine-preventable diseases: poliomyelitis. <www.who.int/ith> (accessed 20.12.03).

WHO, 2014. World Health Organization, WHO: Ebola response roadmap update. <http://www.who.int/csr/disease/ebola/situation-reports/en/> (see also: http://apps.who.int/iris/bitstream/10665/148237/2/roadmapsitrep_14Jan2015_eng?ua = 1).

WHO, 2015. 10 Facts of polio eradication. Available at <http://www.who.int/features/factfiles/polio/en/> (accessed 31.05.16).

WHO, 2016. Zika virus situation report.

WHO, 2016a. Yellow fever in Angola. Available at <http://www.who.int/features/2016/yellow-fever-angola/en/> (accessed 31.05.16).

WHO, 2016b. Angola grapples with worst yellow fever outbreak in 30 years. Available at <http://www.who.int/features/2016/angola-worst-yellow-fever/en/> (accessed 31.05.16).

Wilder-Smith, A., Freedman, D.O., 2003. Confronting the new challenge in travel medicine: SARS. J. Travel Med. 10, 257−258.

Wilder-Smith, A., Quam, M., Sessions, O., Rocklov, J., Liu-Helmersson, J., et al., 2014. The 2012 dengue outbreak in Madeira: exploring the origins. Euro. Surveill. 19, 20718.

Wilder-Smith, A., Leong, W.Y., Lopez, L.F., Amaku, M., Quam, M., et al., 2015. Potential for international spread of wild poliovirus via travelers. BMC Med. 13, 133.

Woodall,, J.P., Yuill,, T.M., 2016. Why is the yellow fever outbreak in Angola a 'threat to the entire world'? Int. J. Infect. Dis. 48, 96−97.

Yang, et al., 2009. The transmissibility and control of pandemic influenza A (H1N1) virus. Science 326, 729−733.

# CHAPTER 12

# Intergovernmental Collaboration in Global Health Informatics

**D. Novillo-Ortiz**

Pan American Health Organization / World Health Organization, Washington, DC, United States

## CONTENTS

## INTRODUCTION

The health of people in the world has experienced noteworthy improvements over the past three decades. However, several critical goals remain unmet regarding global health. Considerable challenges persist in both communicable and noncommunicable diseases; in particular, marked increases in obesity, diabetes, cardiovascular diseases, and cancer, and the continued impact of diseases such as tuberculosis, dengue, and malaria. Suboptimal levels of maternal and child health, insufficient human and infrastructure resources, and wide geographical and cultural differences add further complexity to the situation at a global level. The availability of health services and health-related information varies greatly across communities, geographic areas, and countries, impeding universal access to health services and decreasing the quality of care. These differences are determined by a combination of geographical barriers and other social determinants of health, as well as policy processes and decisions. Viewing the glass as half-full, the spread and uptake of information and communications technologies (ICTs) have the potential to level the playing field by reducing some of these barriers and enabling information-sharing that will assist in equalizing these differences. Many ICT systems and devices, initially expensive and of limited dissemination, have become affordable and are widely used across many levels of society (D'Agostino et al., 2014).

The World Health Organization (WHO) defines eHealth as "cost-effective and secure use of information and communication technologies in support of health and health-related fields" (World Health Organization, 2005). Despite the benefits that health informatics can bring, it is true that today there are significant barriers to its full implementation. A 2013 WHO survey examined obstacles to health informatics implementation. The main obstacles identified are:

- a lack of suitably qualified or experienced professionals to develop and implement health informatics projects;
- inadequate infrastructure to support programs;
- a lack of adequate business models to support broad and sustainable eHealth delivery; and
- a lack of political commitment (WHO, 2014).

A lack of political commitment is one of the main obstacles to implementing health informatics—based systems and services. Political commitment, including intergovernmental collaboration, is key to realizing its full potential, through sustainable, appropriate, and integrated implementation of health informatics initiatives. To move toward a more efficient health sector, it is necessary to establish a governance model that ensures adequate allocation of funds to the sector and facilitates coordination among governmental bodies and between them and other institutions such as universities, industry, and civil society (Novillo-Ortiz, 2015).

In any governance model, intergovernmental collaboration and network play a key role in promoting horizontal network management and collaborative public management. Both concepts are considered as an essential function of public health agencies, which are defined as "mobilizing community partnerships and action to identify and solve health problems." (Centers for Disease Control and Prevention, 2014). In the health informatics context, identifying and solving health problems would be realized through ICT, in particular through two components such as information integration and information sharing.

As an example of the importance of this function, in 2005, the Director General of WHO declared in his opening remarks of a meeting on avian influenza and pandemic human influenza the need to "increase country capacity in surveillance, early detection, diagnosis and reporting of cases—both animal and human." The cost of not being prepared to share information, to coordinate international responses, and to work together is well understood, "If we are unprepared, the next pandemic will cause incalculable human misery" (Jong-wook, 2005; Pardo et al., 2010).

Health informatics initiatives make sense only within a framework of collaboration. Launching a project of this kind in isolation can lead to its failure, which makes intergovernmental collaboration so important. When the political will, available technology, and necessary human and financial resources all exist, intergovernmental collaboration plays a key role in successful implementation of this type of project. The exclusion of a necessary member, the lack of a governance model, or poor communication within the team implementing a health informatics project can mean the failure of an initiative. It should be kept in mind that institutions are represented by people whose will and interests are determining factors in a project's success.

This chapter addresses the intergovernmental collaboration in health informatics including the types of actors needed; the description of the different levels of intergovernmental collaboration at the local, national, and regional levels; some key aspects to consider when initiating intergovernmental collaboration in health informatics and a final conclusion.

## TYPES OF ACTORS IN INTERGOVERNMENTAL COLLABORATION IN GLOBAL HEALTH INFORMATICS

Intergovernmental collaboration facilitates the sustainability of health informatics projects and initiatives. Regardless of the type of government (centralized or decentralized) or how the market is organized (fully or partially regulated, or free market), it is necessary for governance to be based on intergovernmental collaboration that ensures the necessary integration and continuity of processes to obtain the expected results, regardless of the technical, political, and social context.

The following are among the key actors that should be considered regardless of the scope of a health informatics project or initiative (Frenk, 2013), namely: national governments, United Nations system, multilateral development banks (MDBs), philanthropic organizations, global civil society and nongovernmental organizations (NGOs), and private industry. Depending on the scope of the initiative (local, national, or regional), the intensity of stakeholder collaboration and involvement will be substantially different. Table 12.1 illustrates some examples of institutions to take into account within the framework of health informatics.

### National Governments

Governments are responsible for safeguarding the needs of the population and play a key role in the success of health informatics initiatives. In the health sector the two most important entities involved in health informatics, depending on the level of implementation, are the (1) *ministries of health and health centers* and (2) *ministries of information and communications technologies*.

Ministries are responsible for maintaining health systems, while health centers——depending on the level of care——are responsible for directly providing public health services. Together, they are responsible for creating enabling environments and offering new roles for partners, health

**Table 12.1** Types of Actor and Examples

| Types of Actor and Examples |
|---|
| **National Governments** |
| Ministry of Health |
| Ministry of Information Technology and Communications |
| Ministry of Education |
| Academic institutions (Universities, specialized public health schools, and specialized medical informatics schools) |
| **United Nation System** |
| World Health Organization |
| Pan American Health Organization |
| International Telecommunication Union |
| **Multilateral development banks** |
| Regional Development Banks |
| **Philanthropic organizations** |
| Bill and Melinda Gates Foundation |
| Rockefeller Foundation |
| **Global civil society and NGOs** |
| Doctors without Borders (Médecins sans Frontières) |
| Save the Children |
| Professional associations (Physicians, nursing, medical informatics and public health) |
| **Private industry** |
| IT industry |
| Sector private of health |

professionals, authorities, citizens, and others, while identifying disease trends and risk factors and analyzing demographic, social, and health data. As healthcare providers, the health sector has an interest in taking charge of projects to facilitate their success (WHO-ITU, 2012).

The ministries of information technology and communications play a key role in involving private industry and offering guidance and needs analyses for investment and infrastructure. Their responsibilities normally include providing technical assistance to all government sectors, not only the health sector, which means that their integrating vision is essential when implementing health informatics projects designed to have a national scope and impact.

Other government actors include the entities *responsible for the economy and education*, since they are generally responsible for authorizing public spending on health informatics initiatives and play a key role in educating health professionals to meet the high demand for work in the health informatics sector. In most countries, national governments are responsible for approving the academic curriculum, giving the education sector an important role in designing academic plans that serve the needs and demands for health informatics.

Also within the health sector, *academic institutions*, mainly universities, are key players in developing knowledge and evidence that promotes and facilitates decision-making. In particular, health informatics initiatives can be found in medical and nursing schools, which play an important role.

As mentioned earlier, the lack of skilled personnel is one of the main challenges facing the implementation of eHealth in the countries of the Americas. Academic institutions play a key role in filling the need for advanced studies on health informatics. Even more importantly, academic institutions must be allowed to search for the necessary evidence to ensure that investments and decision-making in health informatics are made with certain levels of confidence, safety, and quality.

Within the framework of intergovernmental collaboration, academic institutions are important advisors on conceptual development, methodologies for implementation, and the construction of expert networks to accompany these processes.

Finally, it is important to take into account the perspective of national governments of countries that have *agencies for cooperation and development* whose mission is usually to financially support medium- or low-income countries. These include: the US Agency for International Development (USAID), the Norwegian Agency for Development Cooperation (NORAD), and the Canadian International Development Agency (CIDA), among others.

## United Nations System

The United Nations is an international organization founded in 1945 that currently has 193 Member States.[1] The UN has member programs and special funds, as well as specialized agencies that can be considered strategic partners when establishing intergovernmental collaboration.

The most important member programs and funds involved in health informatics include the United Nations Children's Fund (UNICEF), which offers assistance to children and mothers, and UN Women, which focuses exclusively on gender equity and the empowerment of women. UNICEF has a program for innovation that includes different health informatics projects, as well as an innovation network made up of 14 innovation laboratories around the world. UNICEF's recent partnership with Google to map the Zika virus is a noteworthy initiative.[2]

Three specialized agencies merit special mention: the World Bank, which focuses on poverty reduction and the improvement of living standards worldwide by providing low-interest loans, interest-free credit, and grants to developing countries for education, health, infrastructure, and communications, among other things; the WHO, the public health agency of the United Nations system; and the International Telecommunication Union (ITU), the United Nations specialized agency for information and communication technologies.

Within the framework of health informatics, WHO and ITU are especially relevant since they are two specialized agencies which, in addition to providing technical cooperation in this area, are responsible for developing guidelines for countries and institutions on matters related to health informatics. It is not uncommon to see these two agencies working jointly for greater impact on their objectives. Two specific examples are the development of the National eHealth Strategy Toolkit (WHO-ITU, 2012), which supports the Member States in the development or renewal of national eHealth strategies, and "Be He@lthy, Be Mobile," an mHealth program described in more detail in "Example in India and Egypt: Be He@lthy, Be Mobile" section.

## Multilateral Development Banks

MDBs are institutions that provide financial support and technical assistance to developing countries for economic and social development, through loans with low interests or through grants. Each region of the world has at least one MDBs.

All MDBs support the health sector. For example, the Inter-American Development Bank (IDB), which operates in the "Region of the Americas," has financed a total of 579 health-related projects to date, while the Asian Development Bank has financed 205 and the European Investment Bank has financed 111. Although the use of ICT does not appear in the titles of financed initiatives, practically all projects have at least one health informatics-related component.

MDBs are strategic partners whose intervention can favor the economic viability needed to initiate a health informatics project. Nevertheless, it should be kept in mind that the financial sustainability of an initiative of this kind cannot depend solely on the economic support of MDBs, and that sustainability should always be sought before undertaking a health informatics project. A failure to adequately plan for continuity and for the future of an initiative can mean that a large initial investment may not be completed for lack of financial support.

MDBs also play a key role because they have detailed information that makes it possible to document the outcome, successful or otherwise, of the initiatives and projects that they support financially. This is the main source of information for preparing best practices and lessons learned and for facilitating knowledge exchange between initiatives that pursue the same objectives.

## Philanthropic Organizations

Philanthropic organizations are nonprofit nongovernmental institutions that use their assets to make donations aimed at providing useful social services.

There is a wide range of philanthropic organizations, among which the Bill & Melinda Gates Foundation, the Rockefeller Foundation, and the Wellcome Trust are some of the most important.

In general, institutions of this kind do not finance just any project and typically have their own priorities when providing funding, such as children and health. Nor do they generally finance initiatives for just any country; in particular, it is difficult for developed countries to receive assistance. Africa is a priority region for these institutions.

A positive aspect of philanthropic organizations is that they can play an important role in intergovernmental collaboration by providing seed funding to start health informatics projects. Another important detail that characterizes these institutions is that they tend not to finance the operational phases of projects, which means that they can be approached if the idea is to initiate a project, provided that the required requirements are met and the initiative is aligned with the institution's social agenda. With this in mind, it is important to plan beforehand how the project or the health informatics initiative will be sustained when this initial funding is no longer available.

## Global Civil Society and NGOs

According to the WHO glossary,[3] "civil society is seen as a social sphere separate from both the state and the market. The increasingly accepted understanding of the term civil society organizations (CSOs) is that of nonstate, not-for-profit, voluntary organizations formed by people in that social sphere. This term is used to describe a wide range of organizations, networks, associations, groups and movements that are independent from government and that sometimes come together to advance their common interests through collective action." The term "NGOs," by contrast, "is used to describe nonprofit-making, nonviolent organizations, which seek to influence the policy of governments and international organizations and/or to complement government services (such as health and education)." It is noteworthy that international NGOs include Doctors without Borders (Médecins sans Frontières) and Save the Children.

Within the framework of intergovernmental collaboration, NGOs are sometimes directly in charge of implementing health informatics projects. A practical example is provided in this chapter (see "Example in Africa: Practicing Medicine Without Borders" section), regarding an initiative between Somalia and Kenya that was implemented by Doctors without Borders.

NGOs also include *associations of health informatics professionals*. On many occasions, these associations are governed by an NGO's by-laws, which allows them to continue official relations both with governments and the United Nations system. An example of this is the official relationship between the International Medical Informatics Association and WHO, which collaborate to implement the action plan of the 2005 eHealth Resolution, prepare new eHealth resolutions that address upcoming challenges in the field of interoperability and health data usage, and realize the potential of IT for the strengthening of the Global Healthcare Workforce.

Professional associations are a necessary actor in intergovernmental collaboration mainly because of their capacity to mobilize experts who can provide advice on specific health informatics issues.

## Private Industry

In the context of health informatics private industry refers mainly to the technology sector and the private healthcare sector (e.g., private insurance).

The technology sector plays a key role in the context of intergovernmental collaboration because this is where the technology is developed that makes implementation of health informatics projects viable. Infrastructure development depends to a great extent on this sector whose participation in public projects is usually through public calls for tenders. It is increasingly common for technology companies to make health-related innovations. Apple, Samsung, and other big tech firms are just a few examples of how the industry is trying to provide users with devices that monitor their health, in this case through mobile devices.

Despite suspicions about the interests of private industry, ways exist to establish roles, responsibilities, and limitations for all players in intergovernmental collaboration, making it possible to guarantee transparency between institutions. An example (the epSOS project) is provided in "Example in Europe: European Patient Smart Open Services (epSOS)" section.

Another example can be seen in the private health sector, where insurers have sometimes facilitated the development of client-focused technology such as electronic medical records, electronic prescriptions, and medical appointment reminders. By managing a limited number of patients, it is relatively simple to implement projects of these characteristics.

## LEVELS OF INTERGOVERNMENTAL COLLABORATION

As has been observed, many different actors can play a part in a health informatics project or initiative. Considering a territorial classification, the level of participation in an intergovernmental collaboration will depend on whether the initiative is local, national, or regional in scope.

## Local Level

This level involves a specific locality, municipality, or region (within a country).

The success of a health informatics initiative depends to a great extent on how the project has been designed and planned, and on available resources. Sometimes, even with good planning and adequate provision of resources, there may be a set of external factors—poor adaptation to the introduction of ICT or inadequate change management—that can cause a well-designed project to fail. To know whether a health informatics project is viable, sustainable, and scalable, one option is to implement it at the local or municipal level, as a pilot study.

Although the implementation of a pilot project may appear simpler at the local collaboration level, there are greater risks involved in putting a project into practice for the first time.

Two components of health informatics frequently implemented at this level involve mobile technology and the use of social networks, components that will be used below to illustrate the collaboration at the local level.

### Example in Senegal at District Level: Use of an Interactive Voice Response System to Deliver Refresher Training on Family Planning

In the districts of Meckhé and Tivouane in the Thies region of Senegal, in 2008 and 2009, family planning training was given on contraceptive side effects and misconceptions of health workers. Years later, the Capacity Plus project was implemented to offer a refresher course on this same subject to the same participants who had received the previous training, using mobile technology (a combination of Interactive Voice Response and SMS text messaging). A study endeavored to show the acceptance of this type of project and assess changes in health workers' knowledge of family planning and their perceived ability to counsel clients. This work involved the participation of 20 midwives from 20 health facilities in the area (Gilroy et al., 2015).

Funds for the project were provided by the United States government through the USAID. The project was led by IntraHealth, an NGO working with the government of Senegal for more than 10 years to improve health behaviors and outcomes, including contraceptive use. IntraHealth was entirely responsible for teaching the course in coordination with the Ministry of Health of Senegal, which—through the Department of Reproductive Health and Child Survival—helped prepare the content and select the areas where the project would be implemented. The focal point for the pilot project was also the regional coordinator for reproductive health, which facilitated implementation of the initiative. It was also important to have guidelines from agencies such as WHO and the United Nations Population Fund (UNFPA) to ensure that the content was accurate and up-to-date. Sonatel was the company in charge of providing telephone infrastructure. The Senegalese national ethics committee for health research (*Comité national d'éthique pour la recherche en santé*) played a key role in approval of the study protocol and instruments. In terms of governance and governmental collaboration, Senegal recently created a mHealth Task Force to coordinate all mHealth activities, which will help ensure a strategic vision at the national level going forward. Fig. 12.1 briefly summarizes the role of each of the actors involved in this initiative.

Regarding the results of the project, all the participants said they would retake a course using their mobile phones and 90% reported learning the same amount or more than during an in-person course. Overall, after the training, there was a 17% increase in correct test answers. The greatest costs were associated with the need to contract support for system development and implementation. To scale up this initiative, an identified priority was the need for sufficient time and budgeted financial resources to provide a

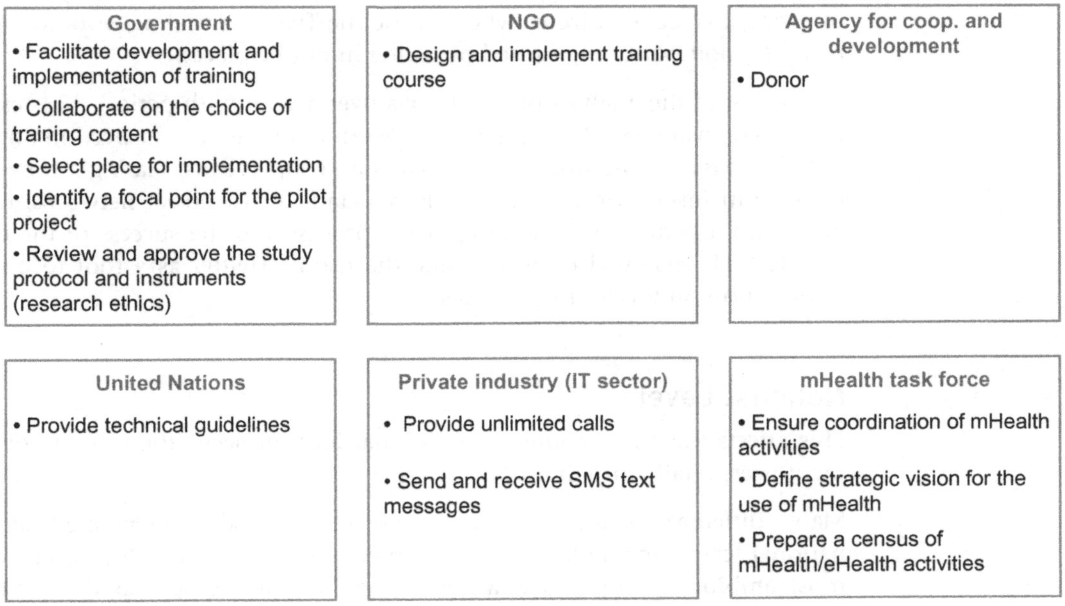

**FIGURE 12.1**

Main role of key actors (Local level).

preliminary period (before the start of official training) for the testing of airtime and contract mechanisms to avoid service provider issues.

### Example in Chicago, Illinois, United States: Using Social Media to Identify Foodborne Illness

If there is a single area in health informatics where traditional intergovernmental collaboration cannot be expected, it involves the use of social networks. The Web is an inexhaustible source of information of interest in the sphere of public health. For example, there was an average of 75 tweets per minute on Zika virus during the week following the WHO announcement of a public health emergency of international concern. Social networks can also play an important role in combatting foodborne disease, a severe health problem with high associated costs. In March 2013 the Chicago Department of Public Health (CDPH) launched the Foodborne Chicago program, with a website (http://www.foodbornechicago.org) aimed at improving food safety in Chicago by identifying and responding to complaints on Twitter about possible foodborne illnesses.

Social networks can provide among the best experiences in collaboration between governments and the public. Authorities in Chicago took this initiative in association with a civic organization (Smart Chicago, devoted to seeking ways to use technology to improve life for residents of Chicago) as a way of giving a voice to citizens, who can use the Twitter social network to report possible food poisoning caused by eating in public places.

As a result of the analysis of 270 tweets over a 10-month period, 133 restaurants were inspected; 21 of them (15.8%) did not pass the inspection and 33 (24.8%) passed conditionally. Inspectors and supervisors, fearing a potential increase in inspections, were initially reluctant to use social networks; however, with a better understanding of the process and the success of the exercise, CDPH personnel came to value the use of Twitter as a tool to obtain information on foodborne diseases.

## National Level

This affects the entire country. In national level projects, the role of government is especially important.

Many different health informatics projects can be implemented at the national level, ranging from the implementation of a mHealth project to the most ambitious project that a government can implement at the national level and requiring the greatest possible intergovernmental collaboration: the adoption of a national eHealth strategy or policy.

### Example From the Americas: Lessons Learned From the Development and Implementation of National eHealth Strategies

The principal objective of national eHealth strategies is to establish a model of governance and sustainability that ensures the continuity of health informatics projects and initiatives over time, regardless of the political and economic context. According to WHO data, 121 countries have an eHealth strategy or policy. In the Region of the Americas 21 countries are now developing or implementing a national eHealth strategy.

Based on WHO and ITU guidelines, we can identify the different groups that would ensure effective intergovernmental collaboration with a governance model that includes all possible actors, namely (see Table 12.2): health sector authorities, a steering committee, a group of experts and stakeholders, and international technical assistance (WHO-ITU, 2012).

In line with this model, major progress on national eHealth strategies is being made in the Region of the Americas. Specific examples can be seen in the experiences of seven countries (Brazil, Chile, Costa Rica, the Dominican Republic, Guatemala, Panama, and Peru). However, while intergovernmental collaboration of this kind can facilitate the success of a national process, there also are certain risks that undermine the likelihood of a good intergovernmental coordination, such as resistance to change, lack of adequate involvement by the principal stakeholders (health professionals), and

**Table 12.2** Key Actors in the Development of a National eHealth Strategy (Vision, Planning, and Evaluation)

| Actors | Responsibilities |
| --- | --- |
| Health sector authorities (high-level decision-making bodies with a political (nontechnical) profile) | Facilitate global guidelines and take charge of obtaining the necessary resources to move forward. Ensure that eHealth priorities are aligned with national health priorities. Establish mechanisms for control and governance of the process; design and establish an independent national eHealth council or committee with stakeholders. |
| Steering committee (mainly people who represent government interests in health and ICT) | Offers direction, makes decisions, and supervises global progress. |
| Group of experts and stakeholders (other involved sectors of government (statistics, legal, office of the president, etc.), potential donors, industry experts, academia, professional associations, civil society, and NGOs, among others) | Provide technical help, guidance, and advice. Potential donors can fully or partially finance certain initiatives. Academia, industry, and NGOs can sometimes implement the project. |
| International technical assistance (international agencies such as WHO or ITU) | Offer high-level guidance to achieve global targets and facilitate the sharing of experiences between countries. Facilitate technical cooperation in project implementation and evaluation. |

fragmentation of the health sector (public and private sectors operating without coordination). Another important risk involves knowing how to make the political authorities understand that the short-term benefits of a governance process are limited and that the best investment they can make is in a process that lays orderly groundwork for eHealth.

### Example in India and Egypt: Be He@lthy, Be Mobile

Government, as a single word, is often assumed to be a single entity. The word itself is misleading: it is in fact a collective noun designating a large number of departments which together guide all decisions taken in the process of managing a state. While a cohesive unit is united by one leader, the different areas themselves are not consistently involved in the work of their counterparts, tending to focus more strongly on their own fields and responsibilities. This can lead to missed opportunities in terms of collaboration and efficient use of financial or knowledge resources.

The field of mHealth—medical and public health practices supported by mobile devices—is by nature highly collaborative. It requires inputs from primary healthcare services, health technology providers, mainstream telecommunications, and regulatory agencies. Activities encompass technical issues, such as service provision, access and data management, through to qualitative issues on patient satisfaction and usability of new services. Meeting this diverse range of needs is significantly easier when departments from the different sectors of public health and ICTs are able to work together (Fig. 12.2).

There are a number of ways interdepartmental collaboration strengthens the programs. The most obvious one is the relevant expertise which each brings to the table. Each department acts as a synecdoche for a far larger network of national organizations and sector-specific contacts, all of whom will contribute to the sustainability of a mHealth program. From the health side, a Ministry of Health will bring all levels of primary, secondary, and tertiary healthcare providers, as well as the patient side including associations and links to existing programs. A Ministry of Communications or ICTs will bring relationships with regulatory agencies, telecom operators, and the broader technology sector. To become embedded as a public health service, mHealth requires a much broader ecosystem than government alone can provide. Drawing on existing connections instead of having to create new ones from scratch can help by leveraging established trust and working relations between the different bodies.

In addition, including several departments can help foster programs which are resilient to political change. Joint ownership means that any change in leadership in one Ministry will not necessarily lead to the demise of the program due to a change in political priorities or delay in implementation

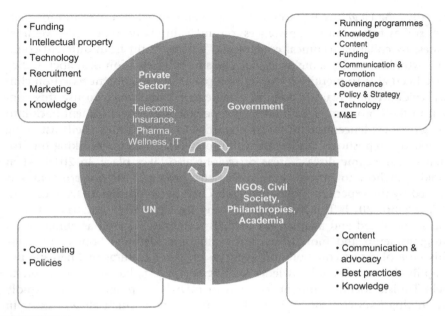

**FIGURE 12.2**

Model of partnership for Be He@lthy, Be Mobile initiative. *WHO-ITU mHealth initiative 2015 Annual Report.*

during the handover period. Shared responsibility can also mean shared resources: a reduction in program costs for each side. Finally it can encourage broader political commitment to a single program.

These theoretical benefits have been demonstrated in two countries implementing national-level mHealth services under a joint mHealth initiative run by the WHO and the ITU. Between 2014 and 2016 India and Egypt both established mHealth programs which are owned jointly by the Ministries of Health and other Ministries concerned with ICTs or eGovernance. In India collaboration with the pan-government digital platform mSeva has guaranteed the mHealth program resources for delivering SMS to the population until 2017, since it is included under the broader national strategy of the Digital India campaign. In Egypt shared ownership of the national program for mDiabetes meant that the Ministry of Science and Communications led the initial work around the technical establishment of an SMS platform and connections with national operators, with the Ministry of Health then taking a lead in service promotion and content.

As an area with limited examples of large-scale projects, mHealth has also been a strong example of how collaboration between governments at the global level can also work to strengthen national programs. The WHO-ITU initiative, Be He@lthy Be Mobile, has been able to act as a platform for sharing successful country experiences on mHealth with other UN Member States. By collating technical content and lessons learnt from different countries working on the same areas, such as smoking cessation or diabetes, they have been able to identify common best practices and content based on clinical evidence and real-life country experiences. Countries who use this content in their programs feed their own experiences and new content back into the global evidence base for future countries to use, creating a self-sustaining knowledge repository. Direct interaction between Ministries working on programs in the same disease areas or regions also takes place. In 2015, when looking at how to introduce an mDiabetes program, the Government was guided by the experiences of Senegal who had been running mDiabetes campaigns since 2014. This extended to using the same SMS database of messages, translated and adapted for cultural suitability, as well as guidance on program setup and monitoring and evaluation. Regional bonds can make this type of direct collaboration even easier, since cultural or linguistic ties amplify the benefits of knowledge-sharing. Egypt, e.g., has since been able to help Tunisia identify content for its own mDiabetes program more rapidly by sharing SMS content from the Egyptian program which already exists in Arabic.

The replicability and intersectoral nature of mHealth has thus highlighted an increasing need for intergovernmental cooperation in the field, but also provided working examples of how this can be managed successfully—along with a rationale for scaling up efforts.

## Regional Level

This affects two or more countries.

The most complex type of intergovernmental collaboration on health informatics involves collaboration between two or more countries. Within the framework of health, relationships between countries are usually defined by reciprocity agreements that allow a person who visits another country to receive medical care at a reduced cost or even free of cost.

The following two initiatives illustrate how to carry out intergovernmental collaboration at the regional level. One is in Europe (exchange of personal health data among health providers in different European countries) and the other in Africa (telemedicine during situations of conflict between countries where there is a shortage of health specialists: Somalia and Kenya).

### *Example in Europe: European Patient Smart Open Services (epSOS)*

Within the framework of the European Union, there is a need for electronic exchange of personal health data among healthcare providers in different

Member States. In practical terms, this collaboration was carried out through the European Patient Smart Open Services (epSOS) project, implemented between 2008 and 2014, with a view to improving the quality and safety of healthcare for European citizens by offering eHealth services. The project consisted in developing summarized patient records (access to important medical data for patient treatment) and an electronic prescription system (see Fig. 12.3).

In terms of intergovernmental collaboration, key aspects of the implementation of this project included the creation of a consortium that brought together national health ministries and other national health authorities, regional health centers, and an industry association. In total, 25 European

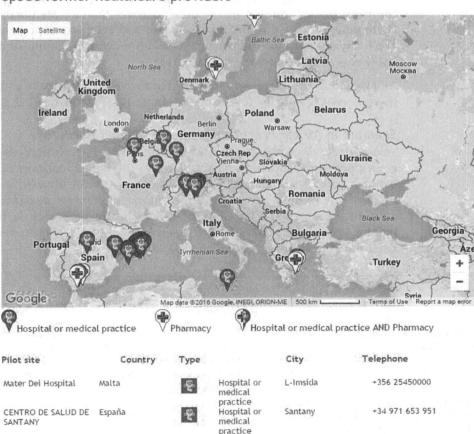

| Pilot site | Country | Type | | City | Telephone |
|---|---|---|---|---|---|
| Mater Dei Hospital | Malta | | Hospital or medical practice | L-Imsida | +356 25450000 |
| CENTRO DE SALUD DE SANTANY | España | | Hospital or medical practice | Santany | +34 971 653 951 |

**FIGURE 12.3**
epSOS former healthcare providers.

countries took part in this project, as well as 45 health agencies and more than 30 companies in the eHealth sector. As for the distribution of roles and responsibilities, the European Union defined "Guidelines on minimum/non-exhaustive patient summary dataset for electronic exchange" and "Guidelines on e-prescriptions dataset for electronic exchange" for drugs and medical devices. Despite these guidelines, protection of personal health data in cross-border healthcare and research settings remains a challenge (Abbing, 2015). For their part, the participating Member States ensured technical and semantic interoperability in the processing and storage of patient data. Finally, companies worked on a pro bono basis and, in return, were offered the opportunity to expand their profesional networks, help drive eHealth in Europe forward, and create business opportunities. Individual vendors could participate by joining the epSOS Industry Team, a consortium of companies represented by Integrating the Healthcare Enterprise (IHE) Europe. To join, a company must, among other things, agree to the Memorandum of Understanding that governs nondisclosure and intellectual property rights. The company must also provide epSOS with a defined amount of expertise, knowledge, and/or services (epSOS, 2014).

An evaluation of this project showed that the major facilitators and barriers to using epSOS were the following (see Table 12.3) (Moharra et al. 2015):

### Example in Africa: Practicing Medicine Without Borders

Telemedicine is another type of health informatics intervention where inter-governmental collaboration has been key in the implementation of projects between countries. In Africa, Somalia and Kenya used telemedicine in 2011 to offer teleconsultations and telementoring to improve pediatric care in areas of armed conflict in Somalia, where health professionals have few training opportunities or adequate supervision and are often unaware of the latest medical advances (Zachariah et al. 2012).

**Table 12.3** Major Facilitators and Barriers to Using epSOS

| Facilitators | Barriers |
| --- | --- |
| Access to patient health information probably improve communication with patients | Concerns around coding of clinical information and medicines (errors in cross coding might lead to inappropriate decisions) |
| Greater knowledge of patients' clinical characteristics and treatment will improve clinical decision-making, i.e.<br>■ Patient summary (e.g., allergies, medical diagnosis)<br>■ ePrescription service: prescribed medication is shown in pharmacist's mother tongue | Lack of integration with national health service information technology systems |
| Simplicity and ease of use (intuitive and user friendly service) | Patient identification and consent process are currently laborious and somewhat time consuming |

Intergovernmental collaboration played a key role in achieving the identified objectives, in this case through Doctors Without Borders, an international humanitarian-aid NGO that runs a district hospital in the Galgadud region of Somalia and that led this intergovernmental collaboration project between Somalia and Kenya. This NGO was in charge of choosing the most appropriate health intervention, in this case telemedicine, in addition to ensuring that the project was implemented according to codes of ethics and the correct guidelines for pediatric management. WHO also played an important role, since its clinical guidelines were also followed to ensure the success of the intervention. Before introducing telemedicine, it was necessary to hold meetings and discussions with another key stakeholder: the community. In a context in which older adults are respected as local authorities by the rest of the population, they played a key role in raising awareness of the importance of the project that was to be initiated with the community, using ICT. The technological industry participated in this initiative through Audiosoft Technologies, which provided technical equipment for the project and ensured secure data transfer. Academia, represented by three senior clinicians, contributed experience to the process. Fig. 12.4 clearly identifies the roles played by the different actors.

The professionals who worked on this project identified telemedicine as useful in improving patient care and, in all cases, the parents of children who received treatment were in favor of this type of intervention. The only barriers identified involved technical problems that were resolved by ensuring dedicated bandwidth. The results of this intervention show how the introduction of telemedicine

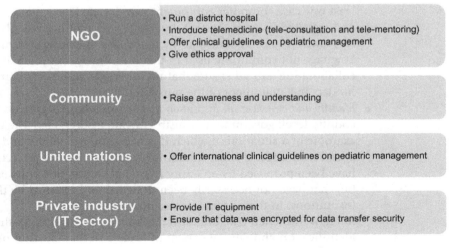

**FIGURE 12.4**
Main roles of key actors (Regional level).

in Somalia, through specialists in Kenya, reduced adverse paediatric ward outcomes (deaths and losses to follow up) by 30% between 2010 and 2011.

## HOW TO ESTABLISH AN INTERGOVERNMENTAL COLLABORATION IN GLOBAL HEALTH INFORMATICS?

Two factors must be taken into account to start this type of collaboration: first, the key considerations and, second, what facilitating elements can generate an enabling environment for creating a context of successful intergovernmental collaboration.

An analysis of these experiences suggests a series of logical steps to establish successful intergovernmental collaboration.

### Key Aspects to Consider When Initiating Intergovernmental Collaboration in Health Informatics

- *Focus intergovernmental collaboration on patients and on the health priorities of the target population.* Intergovernmental collaboration on health informatics should focus on improving peoples' quality of life. Objectives should be very well-defined and promote positive life changes for citizens. To this end, it is essential that promoters of the initiative, both public and private, are clear about the health priorities and needs of the target population. It is essential that health informatics initiatives be developed in response to the country's health priorities, rather than implementing initiatives which, while innovative, may be unnecessary in terms of health priorities. For example, in at least one case computer equipment was donated by a major company and installed in a remote area, but no proper training was planned nor was taken it into account that the machine's operating system was in a language not commonly used in that area. Intergovernmental collaboration should ensure that no other political or economic interest hinders the attainment of the pursued objectives.
- *Identify and empower an institution that can spearhead the process* (at the country level: government ministry, private insurer, etc.) The clearest example of a situation in which it is necessary to identify and empower a lead institution is when initiating health informatics projects that have been proposed by an entity (e.g., a government or private insurer) but that have not necessarily been designed and planned by the same institutions. In these cases it is essential to establish an intergovernmental collaboration mechanism that facilitates the identification and empowerment of leaders in these institutions who can take on the initiative, act as coordinators or focal points, and implement it with full ownership.

- *Identify and involve stakeholders in due time and form.* One of the main activities involved in intergovernmental collaboration is to identify key stakeholders who should be part of a health informatics initiative: universities, technology providers, legal experts, etc. The clearest example is the participation of the teams in charge of actually implementing the project: e.g., in an electronic medical records project, physicians play a key role since they will be the end-users of the tool and their involvement in an initiative of this kind should be planned from the start. Another key role is to review all legal aspects and ensure that the project is ethically and legally viable. In all cases it is advisable to formalize this collaboration through an agreement, whether a formal contract or a memorandum of understanding that clearly states the objectives and type of relationship to be established. As in any formal process, consideration should be given to the timeframes for preparing and signing these agreements, since this can be an obstacle to project implementation. The experience in Africa with telemedicine offered an example of how a delay in the signing of an agreement became a major challenge to carrying out an initiative.
- *Establish a governance model.* As we have seen in previous examples, a governance model can contribute to the development of successful intergovernmental collaboration. It is essential to identify the main stakeholders, establish coordination and communication mechanisms, establish roles and responsibilities, and promote mechanisms for analysis and evaluation that ensure that the governance model works correctly. A governance model will facilitate coordination and proper alignment among the stakeholders in intergovernmental collaboration, and will ensure adequate follow-up in the event that difficulties are encountered in the implementation of an initiative. Intergovernmental collaboration can be understood as a collaboration agreement between institutions working to achieve a single objective; in the event of structural changes in any of the participating stakeholders, a governance model can help ensure the institutional sustainability required in health informatics projects, which always involve a large number of actors.

Furthermore, a series of factors facilitate a context of successful intergovernmental collaboration.

### Actions Needed to Create an Adequate Environment and Develop Elements That Facilitate Intergovernmental Collaboration in Health Informatics

- *Favor cross-cutting action and intersectoral dialog (political-social environment).* In a setting in which health is the most important component, it is important to share that prominence with other key sectors, like technology, to allow for adequate inclusion and facilitate the necessary cross-cutting work so that all participants are the main actors in their respective roles. A lack of integration of the health sector, or its isolation, will lead to a lack of adequate intergovernmental collaboration.

■ *Use a portability-based environment (virtual environment).* Portability and online presence make today's collaboration and communication between institutions much more nimble and straightforward than they were 10 or 15 years ago. Since willingness to engage in interagency dialog is key for developing intergovernmental collaboration, portability will facilitate this task, but it should never fully replace ongoing, in-person contact among the different actors.

■ *Ensure financial sustainability (economic environment).* As in every initiative, financial resources will largely facilitate the project's sustainability and continuity. As examined previously, there are institutions (e.g. MDBs or philanthropic organizations) that can support the launch of such a project. Nevertheless, adequate planning is essential, and a financial plan must be in place in which the institution implementing the project or initiative has established a financial viability plan for maintaining and potentially scaling up the project.

■ *Review and adapt the legal framework (legal environment).* An exhaustive analysis of the legal framework—not just for health sector—will help to make the necessary adaptations to reduce the challenges and possible setbacks involved in implementing health informatics projects, where the exchange of personal data and data privacy are among the major challenges facing countries working on health informatics.

■ *Follow international technical guidelines and past experiences (technical environment).* To promote an environment conducive to intergovernmental collaboration, it is also necessary to have the involvement and technical assistance of institutions that provide technical guidance (in the case of health informatics, agencies in the United Nations system, specifically the World Health Organization and the ITU, play a prominent role), together with other institutions—including other countries—that have implemented such initiatives in the past and can help shed light on the main challenges that might arise and how to overcome them.

■ *Identify the available evidence supporting decision-making (scientific environment).* Incorporating available evidence when initiating an intergovernmental collaboration process will help support decision-making during implementation of the initiative. There will certainly be times when questions arise about the path to take, and existing evidence will be an important component to take into account.

# CONCLUSIONS

There are three factors that illustrate the usefulness and benefits of intergovernmental collaboration in health informatics: first, ICTs are an essential component for improving people's health; second, evidence indicates that a lack of business models and lack of political commitment are two of the main barriers to moving forward in health informatics projects; and third, the case studies analyzed in this chapter show that no health informatics project involves just one institution, and that a diverse set of actors is actively involved in this series of initiatives.

These three factors illustrate the need to work through intergovernmental collaboration, facilitating communication and promoting the sustainability of the projects that involve key actors, including national governments, United Nations agencies, MDBs, philanthropic organizations, civil society, nonprofit governmental organizations, and private industry. Each actor's role will be different depending on the scope of the initiative. For example, while international organizations typically have a relevant role in technical assistance and providing technical guidelines, NGOs and donors may have various functions ranging from technical assistance and financing to the actual implementation of an initiative.

This is the context, regardless of the scope of the initiative (local, national, or regional) or type of government (centralized or decentralized), or how the technology and health market is organized (fully or partially regulated, or free market). Specifically in the health sector, where fragmentation exists intergovernmental collaboration has particular value added, since it can facilitate greater integration among health informatics projects and initiatives that can directly benefit the sector.

The benefits of intergovernmental collaboration in health informatics are evident in that they help achieve results; nevertheless, major challenges remain, most notably a potential lack of understanding among the different actors involved in such projects. A lack of intergovernmental collaboration can result in isolated initiatives that may function but would be difficult to scale up. Establishing a governance model can certainly prevent this and guarantee the communication and information sharing needed to facilitate the implementation of initiatives.

Finally, it should be kept in mind that if a health informatics project or initiative is successful, no one will mention the importance of having worked through intergovernmental collaboration whereas, if the expected results are not achieved, all the analyses will ultimately conclude that the project did not work due to a lack of intergovernmental collaboration.

The five separate fingers are five independent units. Close them and the fist multiplies strength. This is organization.

**James Cash Penney.**

## Acknowledgments

We would like to thank Ms. Susannah Robinson (WHO) and Dr. Sameer Pujari (WHO) for their support in the preparation of "Example in India and Egypt: Be Healthy Be Mobile." Special thanks also go to the rest of the team that makes the Be He@lthy Be Mobile initiative possible (Dr. Vinayak Prasad, Ms. Gini Arnold, Mr. Per Hasvold, Mr. Hani Eskandar, Ms. Surabhi Joshi, Ms. Allison Goldstein, and Ms. Stephanie Meagher).

**Disclaimer** The author is staff member of the Pan American Health Organization / World Health Organization (PAHO/WHO) and is himself alone responsible for the views expressed in the chapter, which do not necessarily represent the views, decisions, or policies of the PAHO/WHO

## End Notes

1. United Nations. Funds, Programmes, Specialized Agencies and Others. Available from: http://www.un.org/en/sections/about-un/funds-programmes-specialized-agencies-and-others/index.html
2. Google. Providing support to combat Zika in Brazil and beyond. March 3, 2016.https://googleblog.blogspot.com/2016/03/providing-support-to-combat-zika-in.html
3. World Health Organization. Civil society. Available from: http://www.who.int/trade/glossary/story006/en/

## References

Abbing, H.D., 2015. Eu cross-border healthcare and health law. Eur. J. Health Law 22 (1), 1–12.

Centers for Disease Control and Prevention, May 29, 2014. National Public Health Performance Standards. The Public Health System and the 10 Essential Public Health Services. Available at: <http://www.cdc.gov/nphpsp/essentialservices.html> (accessed 29.03.16).

D'Agostino, M., et al., 2014. eHealth initiatives to transform health in the Region of the Americas. Pan Am. J. Public Health. Available at: <http://bit.ly/1sNZGDk>. (accessed 29.03.16).

European Patient Smart Open Services (epSOS), 2014. Available at: <www.epsos.eu> (accessed 12.04.16).

Frenk, J., Moon, S., 2013. Governance challenges in global health. N. Engl. J. Med. 368 (10), 936–942, <http://dx.doi.org/10.1056/NEJMra1109339>.

Gilroy, K., et al., 2015. Use of an interactive voice response system to deliver refresher training in Senegal. Findings from Pilot Implementation and Assessment. IntraHealth, Washington, DC. Available at: <http://www.intrahealth.org/page/use-of-an-interactive-voice-response-system-to-deliver-refresher-training-in-senegal> (accessed 16.04.16).

Jong-wook, L., 2005. Meeting on Avian Influenza and Pandemic Human Influenza: Opening Remarks. World Health Organization, Geneva, Switzerland.

Moharra, M., Almazán, C., Decool, M., Nilsson, A.L., Allegretti, N., Seven, M., 2015. Implementation of a cross-border health service: physician and pharmacists' opinions from the epSOS project. Fam. Pract 32 (5), 564–567.

Novillo-Ortiz, D., 2015. Estrategias nacionales de eSalud en la región de las Américas: avances y desafíos. Rev. MedUNAB 17 (3), S26–S30, Suppl. 2015.

Pardo, T.A., Gil-Garcia, J.R., Luna-Reyes, L.F., 2010. Collaborative governance and cross-boundary information sharing: envisioning a networked and it-enabled public administration. In: O'Leary, Van Slyke, D., Kim, S. (Eds.), The Future of Public Administration, Public Management and Public Service Around the World: The Minnowbrook Perspective. Georgetown University Press, Washington, D.C.

World Health Organization, 2005. Resolution WHA58.28. eHealth. In: Proceedings of the Fifty-eighth world Health Assembly, Geneva, 16–25 May 2005. WHO, Geneva. (WHA58/2005/REC/1). Available from: <apps.who.int/gb/or/e/e_wha58r1.html> (accessed 29.03.16).

World Health Organization, International Telecommunication Union, 2012. National e-Health strategy toolkit. Geneva.

World Health Organization, International Telecommunication Union, March 2014. eHealth and innovation in women's and children's health: a baseline review based on the findings of the 2013 survey of CoIA countries by the WHO Global Observatory for eHealth. Available from: <www.who.int/goe/publications/baseline_fullreport/en/> (accessed 29.03.16).

Zachariah, R., et al., 2012. Practicing medicine without borders: tele-consultations and tele-mentoring for improving paediatric care in a conflict setting in Somalia?. Trop. Med. Int. Health 17 (9), 1156–1162, <http://dx.doi.org/10.1111/j.1365-3156.2012.03047.x>. Epub 29.07.12.

# Author Index

*Note*: Page numbers followed by "*f*" and "*t*" refer to figures and tables, respectively.

# Subject Index

*Note*: Page numbers followed by "*f*" and "*t*" refer to figures and tables, respectively.

## A

Academic Forum (AMIA), 3
Academic institutions, 268
Academic Model Providing Access to
    Healthcare (AMPATH), 209
Accessibility, 8
ACR. *See* American College of
    Radiology (ACR)
Act management (AM), 132–133
Adaptation, 8
Admission-discharge transfer (ADT)
    function of CIS, 132
Africa
    EHR in, 59–60
    global disease burden, 1–2
    health workforce, 1–2
    telemedicine in, 280–282
Aging population, 1, 8
    healthcare system and, 219
American College of Physicians, 82
American College of Radiology
    (ACR), 95
American Health Information
    Management Association, 96
American Medical Informatics
    Association (AMIA), 3
Analytics. *See* Data analytics
Anatomical-Therapeutic-Chemical
    (ATC) classification, 99
Ancestor-Descendant-Identity (ADI),
    119
Apelon, 118
Application Programming Interfaces
    (API), 32
Appointment and resource
    scheduling (ARS), 133
Apps (mobile), 6, 159
    WHO classification categories, 158

Argentina, 140
Asia-Pacific region, EHR in, 59
Associations of health informatics
    professionals, 271
ASTM International, 95

## B

Bangladesh, mHealth in, 72
Be He@lthy Be Mobile, 277–278
Big data, 198–199, 205–206
Bill & Melinda Gates Foundation,
    270
Biochemical test sensors, 159
Biochips, 181, 190–191
Bioethics, 233–234
    infectious diseases, 235–236
    information technology, 234–235
Biomedical informatics (BMI), 3
Biomedical Translational Research
    information System (BTRIS),
    111
    controlled terminologies,
        117–120
    data access authorization,
        120–122
    data model, 114–117, 115*f*
    data reuse notification, 121–122
    data sources, 111–113, 112*t*
    data storage in, 118–119
    deidentified data, 114
        queries, 122
        retrieval of, 120
    global health and, 124–126
    identified data
        queries, 122
        retrieval of, 120
    retrieving data based on protocol,
        113–114

    terminology lookup in, 123–124
    tracking clinical protocol
        enrollment, 113
    user interface components, 122
Biometrics, 198
BMI. *See* Biomedical informatics
    (BMI)
BOADICEA, 207–208
Brazil
    EHR in, 60
    mHealth in, 72
    National Telecommunications
        Agency, 156
BYOD + K, 158

## C

CaBIG, 6–7
CaGrid, 6–7
Canada, EHR in, 57–58
Canadian Institutes for Health
    Research (CIHR), 20–21
Canadian International Development
    Agency (CIDA), 268
Capacity Plus project, Senegal,
    272–274
CCD. *See* Continuity of Care
    Document (CCD)
CDISC. *See* Clinical Data Interchange
    Standards Consortium
    (CDISC)
CDPH. *See* Chicago Department of
    Public Health (CDPH)
CEN. *See* European Committee for
    Standardization (CEN)
Center for Advancing Health,
    223
Chicago, United States, 274
    Foodborne Chicago program, 274

Printed in the United States
By Bookmasters